T0235285

Geographies of the 2020 U.S. Presidential Election

This timely, insightful and expert-led volume interprets the 2020 U.S. Presidential Election from a geographical standpoint, with a focus on its spatial dimensions.

With contributions from leading thinkers, this book highlights the unique circumstances of the election, including the Covid pandemic and a president who falsely alleged that it was a massive fraud, particularly after he lost. The volume offers an introduction and 11 chapters that examine the run-up to the election, the motivations of Trump supporters, the election results themselves, case studies of the battleground states of Wisconsin and Georgia, and the chaotic aftermath. Accompanied with an engaging plethora of figures providing a visual demonstration of data trends, both national and local case studies are considered throughout this book, as well as right-wing radicalization, the role of Cuban-Americans, race, and threats to American democracy.

This book is an ideal study companion for faculty and graduate students in fields, including geography and political science, sociology, American studies, media studies and urban planning, as well as those with an interest in U.S. politics more generally.

Barney Warf is a Professor of Geography at the University of Kansas. Much of his research concerns producer services and telecommunications, particularly the geographies of the internet, including fiber optics, the digital divide, e-government, and internet censorship. He views these topics through the lens of political economy and social theory. He also maintains an active interest in political geography, including elections, voting technologies, and the U.S. electoral college. Currently, he serves as editor of *GeoJournal*, editor-in-chief for geography for Oxford Bibliographies On-Line, and edits a series of geography texts. His teaching interests include urban and economic geography, the history of geographic thought, globalization, and contemporary social theory.

John Heppen is a Professor of Geography at the University of Wisconsin River Falls. With David Beard, he has written on professional wrestling for *Political Landscapes in the Age of Donald Trump*, for the *Popular Culture Studies Journal*, for *Iconic Sports Venues: Persuasion in Public Spaces*, and for *Sports Fans, Identity, and Socialization: Exploring the Fandemonium*. They present research at the annual WrestlePosium, the conference of the Professional Wrestling Studies Association. He is also one of the co-editors of *Atlas of the 2016 Elections* and *Atlas of the 2020 Elections*.

Routledge Research in Place, Space and Politics
In memory of founding series editor Professor Clive Barnett,
University of Exeter, UK

This series offers a forum for original and innovative research that explores the changing geographies of political life. The series engages with a series of key debates about innovative political forms and addresses key concepts of political analysis such as scale, territory and public space. It brings into focus emerging interdisciplinary conversations about the spaces through which power is exercised, legitimized and contested. Titles within the series range from empirical investigations to theoretical engagements and authors comprise of scholars working in overlapping fields including political geography, political theory, development studies, political sociology, international relations and urban politics.

Postsecular Geographies
Re-envisioning Politics, Subjectivity and Ethics
Paul Cloke, Christopher Baker, Callum Sutherland, and Andrew Williams

Oil, Culture and the Petrostate
How Territory, Bureaucratic Power and Culture Coalesce in the Venezuelan Petrostate
Penélope Plaza Azuaje

Migration in Performance
Crossing the Colonial Present
Caleb Johnston and Geraldine Pratt

Spaces of Tolerance
Changing Geographies and Philosophies of Religion in Today's Europe
Edited by Luiza Bialasiewicz and Valentina Gentile

Geographies of the 2020 US Presidential Election
Edited by Barney Warf and John Heppen

For more information about this series, please visit: www.routledge.com/series/PSP

Geographies of the 2020 U.S. Presidential Election

Edited by Barney Warf and John Heppen

Routledge
Taylor & Francis Group

LONDON AND NEW YORK

First published 2023
by Routledge
4 Park Square, Milton Park, Abingdon, Oxon OX14 4RN

and by Routledge
605 Third Avenue, New York, NY 10158

*Routledge is an imprint of the Taylor & Francis Group,
an informa business*

© 2023 selection and editorial matter, Barney Warf and John Heppen;
individual chapters, the contributors

The right of Barney Warf and John Heppen to be identified as the
authors of the editorial material, and of the authors for their
individual chapters, has been asserted in accordance with sections
77 and 78 of the Copyright, Designs and Patents Act 1988.

All rights reserved. No part of this book may be reprinted or
reproduced or utilised in any form or by any electronic, mechanical,
or other means, now known or hereafter invented, including
photocopying and recording, or in any information storage or
retrieval system, without permission in writing from the publishers.

Trademark notice: Product or corporate names may be trademarks or
registered trademarks, and are used only for identification and
explanation without intent to infringe.

British Library Cataloguing-in-Publication Data
A catalogue record for this book is available from the British Library

Library of Congress Cataloging-in-Publication Data
Names: Warf, Barney, 1956- editor. | Heppen, John, editor.
Title: Geographies of the 2020 US presidential election / edited by
Barney Warf and John Heppen.
Other titles: Geographies of the 2020 United States presidential
election
Description: First Edition. | New York : Routledge, 2023. |
Series: Routledge Research in Place, Space and Politics |
Includes bibliographical references and index.
Identifiers: LCCN 2022012824 (print) | LCCN 2022012825 (ebook) |
ISBN 9781032197821 (Hardback) | ISBN 9781032197838 (Paperback) |
ISBN 9781003260837 (eBook)
Subjects: LCSH: Electoral geography--United States. | Presidents--
United States--Election--2020. | Elections--United States--Statistics. |
Voting--United States. | Political campaigns--United States. |
COVID-19 Pandemic, 2020--Influence. | Republican Party (U.S. : 1854-)
Classification: LCC JK1976 .G46 2023 (print) | LCC JK1976 (ebook) |
DDC 324.973/0933--dc23/eng/20220803
LC record available at https://lccn.loc.gov/2022012824
LC ebook record available at https://lccn.loc.gov/2022012825

ISBN: 978-1-032-19782-1 (hbk)
ISBN: 978-1-032-19783-8 (pbk)
ISBN: 978-1-003-26083-7 (ebk)

DOI: 10.4324/9781003260837

Typeset in Times New Roman
by SPi Technologies India Pvt Ltd (Straive)

Contents

Figures

Tables

Contributors

David Beard is a Professor of Rhetoric at the University of Minnesota Duluth. With John Heppen, he has written on professional wrestling for *Political Landscapes in the Age of Donald Trump*, for the *Popular Culture Studies Journal*, for *Iconic Sports Venues: Persuasion in Public Spaces*, and for *Sports Fans, Identity, and Socialization: Exploring the Fandemonium*. They present research at the annual WrestlePosium, the conference of the Professional Wrestling Studies Association.

Fiona M. Davidson has a PhD in Geography from the University of Nebraska-Lincoln and has worked as a political geographer at the University of Arkansas for almost 30 years. She has a long-standing research interest in European politics and nationalism, especially in the United Kingdom, and has spent the last decade writing on elections in the United States. As one of the editors of the *Atlas of the US Election* (2008, 2012, 2016 and 2020), she has developed a significant research interest in the demographics of U.S. elections, with particular focus on the voting and representation of under-represented groups.

Adam S. Dohrenwend is a PhD student at Louisiana State University's Department of Geography and Anthropology. His dissertation research focuses on electoral politics, industrialization, and environmental (in-)justice in south Louisiana – an electoral and industrial political ecology. His previous research has examined socioenvironmental impacts of yerba mate cultivation in northeastern Argentina. He earned his BA in geography at SUNY Geneseo in May 2017 and his MA in geography at the University of Kansas in May 2019. He expects to conclude his PhD studies at Louisiana State University in 2024.

Jamey Essex is Professor in the Department of Political Science at the University of Windsor, in Windsor, Ontario, Canada. He has research and teaching interests in political geography and geopolitics, especially as these relate to development aid and diplomacy, international political economy and international relations, and agriculture and food issues. His research on official development assistance, specifically those state institutions that handle donor countries' aid and development policies, has examined how these institutions frame and intervene in processes such as agricultural development, national security, and governance, and how this relates to their internal structures and external relations. More recent research on diplomacy has examined how expertise, gender, institutional culture, and place matter for the geographies of diplomatic practice, with particular attention to changes at Global Affairs Canada. He has published in *The Annals of the American Association of Geographers*, *Geopolitics*, *The Canadian Geographer*, *Political Geography*, and the *International Feminist Journal of Politics*.

Dr. Kenneth French is an Associate Professor in the Department of Geography and Anthropology and the Director for the Center for Ethnic Studies at the University of Wisconsin-Parkside. As an urban geographer, his research interests pertain to ethnic residential segregation, ethnic enclaves, electoral geography, and the geography of rap music.

Jakob Hanschu is a PhD student in sociocultural anthropology and a Harvey Graduate Fellow in American Culture Studies at Washington University in St. Louis. He is a Fulbright alumnus and holds an MA in Critical Theory and Politics from the University of Nottingham. His research focuses on the political economy and historical geography of industrial agriculture, the environmental governance of multi-scalar environmental phenomena, and the material politics of infrastructure in the U.S. Midwest.

John Paul Henry is a PhD candidate and political geographer at the University of Kansas, where he studies surveillance, authoritarianism, and geopolitics. His dissertation concerns Cuban state surveillance and the activist practice of resistance and seeing from below, called sousveillance. Henry is interested in visual culture as a regime of truth and visuality's role in geopolitics. Henry's dissertation research has been recognized and supported by the Harry S. Truman Good Neighbor Award Foundation and he is the recipient of the University of Kansas College of Liberal 2020 Arts Social Justice Award.

John Heppen is a Professor of Geography at the University of Wisconsin River Falls. With David Beard, he has written on professional wrestling for *Political Landscapes in the Age of Donald Trump*, for the *Popular Culture Studies Journal*, for *Iconic Sports Venues: Persuasion in Public Spaces*, and for *Sports Fans, Identity, and Socialization: Exploring the Fandemonium*. They present research at the annual WrestlePosium, the conference of the Professional Wrestling Studies Association. He is also one of the co-editors of *Atlas of the 2016 Elections* and *Atlas of the 2020 Elections*.

Laurie M. Johnson is Professor of Political Science and Director of the Primary Texts Certificate at Kansas State University. She is the author of seven books and numerous book chapters and articles. Most of her work has involved developing a thorough understanding and critique of classical liberal theory, and includes works on Hobbes, Locke, Rousseau and Tocqueville. Her most recent book, *Ideological Possession and the Rise of the New Right: The Political Thought of Carl Jung*, was published in 2019 by Routledge. She is currently working on a new book, *The Gap in God's Country: Towards Repairing Our Rural/Urban Divide*.

Shaun J. Johnson is an MA student in Geography at the University of Kansas whose research focuses on American political and electoral geographies.

Kimberly Johnson Maier is an instructor in Geography at South Dakota State University. She is currently finishing her PhD in Geography from Oklahoma State University with a focus on historical geography and tourism. She has research interests in women's history and U.S. elections and is one of the editors of the *Atlas of the 2020 US Election*.

Fred M. Shelley is Professor Emeritus of Geography and Environmental Sustainability at the University of Oklahoma. His research interests include electoral and political geography, demography, and the geography of the United States. He is a co-editor of, and a contributor to, the atlases of the 2008, 2012, 2016, and 2020 elections, and he is the author of numerous other publications in his areas of interest and expertise.

Abraham Stephenson is a writer and educator in Cuba. Stephenson previously worked at the University of Havana but left due to institutional political indoctrination. Stephenson's pedagogical interests include geography, history, culture, and the current situation in Cuba. Stephenson's writing interests are diverse, covering the scientific, social, historical-cultural, and philosophical literature.

Barney Warf is a Professor of Geography at the University of Kansas. Much of his research concerns producer services and telecommunications, particularly the geographies of the internet, including fiber optics, the digital divide, e-government, and internet censorship. He views these topics through the lens of political economy and social theory. He also maintains an active interest in political geography, including elections, voting technologies, and the U.S. electoral college. Currently, he serves as editor of *GeoJournal*, editor-in-chief for geography for Oxford Bibliographies On-Line, and edits a series of geography texts. His teaching interests include urban and economic geography, the history of geographic thought, globalization, and contemporary social theory.

Ryan Weichelt is a Professor in the Geography and Anthropology Department at the University of Wisconsin-Eau Claire. He authored several articles and chapters focusing on the electoral geography of Wisconsin, redistricting, and second home ownership in Wisconsin's Northwoods. He was an editor and contributor to the *2020 Atlas of Elections* and the *2016 Atlas of Elections*.

1 Introduction

The 2020 Presidential Election in Context

Barney Warf and John Heppen

The 2020 U.S. presidential election was a momentous moment in contemporary American political history that reflected a deeply divided and polarized country. As Paulson (2021, p. 105) notes, "It was an election year like no other." The incumbent president, Donald Trump, faced a stiff challenge from Democratic nominee Joe Biden, whose vice-presidential candidate was the first woman (and notably a woman of color, Kamala Harris) to be elected to the office. For many voters, the election was a referendum on Trump and his four tumultuous years in office. After a long campaign, Biden won 306 votes in the Electoral College versus 232 for Trump (Figure 1.1), identical to Trump's victory in 2016 (which he called a "landslide"). Turnout, at 67%, was the highest in the U.S. in a century; more than 158 million people voted. Roughly 81 million voters chose Joe Biden, the largest number of votes ever garnered by a candidate, while 74 million chose Trump, an increase of 10 million over his surprising victory in 2016. (Trump lost the popular vote in both elections, but won the Electoral College in 2016). In many states the margins of victory were razor-thin. Democrats flipped five states carried by Trump in 2016; they restored their "Blue Wall" in the Midwest (Wisconsin, Michigan) and in Pennsylvania, and flipped two traditionally Republican-leaning states (Arizona and Georgia). Notably, had 43,000 voters in three states (Wisconsin, Georgia, and Arizona) switched their votes, Trump would have been re-elected.

The harshness of rhetoric exhibited was unmatched in recent U.S. history, with accusations of socialism, fascism, conspiracies, and more. Threats of violence loomed. The country was engulfed in tensions over police oppression of minorities, as indicated by the Black Lives Matter movement and racial unrest in many cities sparked by the murder of George Floyd in Minneapolis. Moreover, the election occurred in the midst of the Covid pandemic, which had killed hundreds of thousands of people (Baccini et al. 2021). Trump was heavily criticized for mismanaging the pandemic, which likely cost him the election. Trump himself was hospitalized with Covid-19 in October after he spent months making misleading statements about the virus, downplayed its seriousness, suggested people inject household antiseptics into their bodies to fight the virus, and refused to wear masks in public. The pandemic encouraged historically high levels of mail-in votes and an

DOI: 10.4324/9781003260837-1

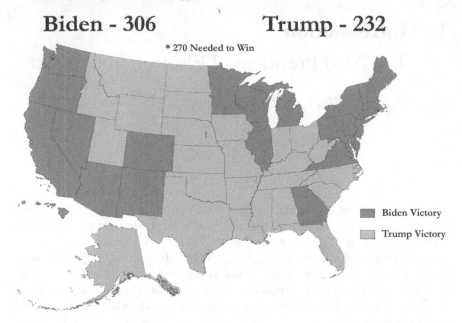

Biden - 306   **Trump - 232**

* 270 Needed to Win

Biden Victory
Trump Victory

Figure 1.1 Electoral College Results of the 2020 Election.

Source: Wikicommons, https://commons.wikimedia.org/wiki/File:2020_Presidential_Election.png

economy mired in recession. Of the 158 million total votes cast, more than 100 million were cast before election day via early voting or mail-in votes. The pandemic also led to a shortage of election workers and reduced number of polling places. The large number of mail-in ballots led to delays in vote counting, fueling allegations of fraud.

Upon Trump's defeat, he refused to concede defeat, the first American president to do so. Trump and his allies subsequently waged an unprecedented war on the electoral process, promoting the lie that Biden won unfairly through election fraud and seeking to overturn the popular and Electoral College votes through a series of lawsuits that uniformly failed due to lack of evidence. Despite Republican allegations, there was no evidence of widespread voter fraud (Eggers et al. 2021). Some of Trump's acolytes engaged in wild conspiracy theories. The period after the election was plagued by unfounded claims of fraud and demands for vote recounts. Urban voters and minority voters were often subject to such attacks. Such moves raised fears about the future of democracy in America (Levitsky and Ziblatt 2018). In the culmination of his attempts to overturn the election results, Trump and followers incited a riot that led to a violent assault on the U.S. Capitol building and deaths of police officers and even one rioter on January 6, 2021, leading to Trump's second impeachment but eventual acquittal. Ultimately, Biden and Harris were inaugurated on January 20, 2021.

The election reflected a confluence of factors that made it particularly significant and turbulent. In the U.S., the rising tide of right-wing populism emboldened Trump, fellow Republicans, and other reactionaries. Blue-collar and rural voters flocked to the Republican Party in droves. The election amplified racial tensions between Whites and Blacks, but also saw many Latinos turn to the Republican Party. Conversely, many voters were deeply repelled by Trump's dishonesty, insults, racism, misogyny, and boasts. Women and suburbanites rejected him in large numbers, providing Biden with the margin of victory. Some Republicans who had backed Trump in 2016 turned away from him in 2020. Trump became the first president since 1992 to be defeated in his re-election bid, only the third since World War II, and the first since Herbert Hoover to lose the White House, Senate, and House of Representatives during his term.

Democratic control over the House, which they took in 2018, was very narrow, with a margin of only five votes. In the Senate, special elections in Georgia were won by two Democratic candidates, tying the two parties at 50 Senators each, although the vice-president's authority to break ties gave Democrats a 51–50 edge. The influence of campaign donations was unprecedented: the campaigns and election cost an estimated $14 billion. Finally, suspicions of foreign interference such as hacking by Russia and Iran lurked in the background.

A geographical perspective on the election offers numerous insights. Support for the two parties was highly uneven across the country (Figure 1.2). Support for Republicans was highest in the South and Midwest, traditionally conservative areas, and rural regions, while Democrats won many voters along the coasts, in large cities, and in areas with large African-American

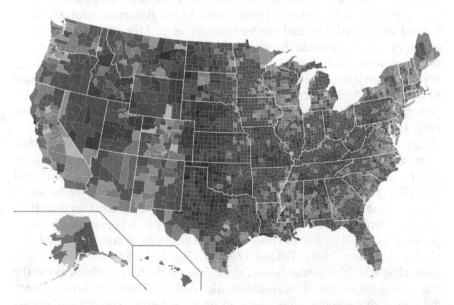

Figure 1.2 Democratic and Republican Popular Vote Victories by County in the 2020 Election.

Source: Wikicommons, https://commons.wikimedia.org/wiki/File:2020_United_States_presidential_election_results_map_by_county.svg

populations. The urban–rural divide was stronger than ever: Biden won 60% of votes in metropolitan areas, while Trump won 57% of votes in non-metropolitan ones. Trump was especially popular with non-college-educated White voters in rural areas; in some rural Texas counties, he won 96% of the votes. African-Americans remained strongly Democratic in their preferences, while Protestants and Evangelicals fervently supported Trump. Republicans made important inroads in Latino communities in south Texas and southern Florida. However, in many places, Trump underperformed relative to other Republican candidates. Suburban voters defected to Democrats *en masse*. The gender gap, which long favored Democrats, remained firmly in place: Biden won 57% of women's votes, while Trump won 53% of men's. In addition, the uneven spatiality of the coronavirus pandemic differentially affected turnout and rates of mail-in voting. Trump and Republicans strenuously objected to mail-in votes, even though the GOP had relied on them for years, on the grounds that they facilitated election fraud. Finally, because geography is concerned about place as well as space, the volume offers several case studies that reveal how broad political dynamics play out in unique circumstances.

Electoral Geography Past and Present

The current volume arises from and contributes to a long and rich tradition in electoral geography stretching back to the 1950s (Johnston 2005; Johnston et al. 1990; Prescott 1959; Warf and Leib 2011). The subdiscipline has been largely focused on the Anglophone world, with numerous British contributions (e.g., Johnston and Pattie 2006). In the American context, many focused on redistricting and gerrymandering (e.g., Morrill 1981). A steady stream of works examined the spatiality of elections, including turnout rates, voting patterns, and changes over time (Archer 1988; Archer and Shelley 1988). Research involved case studies of those elections held in 1976 (Swauger 1980), 1984 (Archer et al. 1985), and 2004 (Wing and Walker 2010). Geographers have excelled at making atlases of elections too (e.g., Martis 1982, 1989; Martis et al. 2014).

Electoral geography has gone through many twists and turns over time, from its empiricist and positivist roots to more recent and more nuanced understandings drawn from political economy and social theory. Warf and Leib (2011) held out hope for a renaissance in the field, propelled by social theory on the one hand and geographical information systems on the other. Electoral geographers moved from mapping votes to a concern with social power (Forest 2018). Works concerned with elections in Italy (Agnew 1995; Shin and Agnew 2008), Poland (Zarycki 2015), Turkey (Bekaroğlu and Osmanbaşoğlu 2021), and Bosnia (Reményi et al. 2021) offered hope the field would break out of its Anglophonic confines. Some scholars explored the notion that elections are arenas in which subjects express their preferences within structural constraints, ranging from the local scale to the world system (Agnew 1996; Johnston and Pattie 2006). Yet others, such as Warf (2009), examined the Electoral College and the spatial biases it generates.

Donald Trump, as flamboyant and contentious a figure as there ever was, merits geographical attention too (Warf 2021).

Outline of Chapters

This volume scrutinizes the geography of the 2020 election from a variety of conceptual angles and spatial scales. Some are concerned with national processes, while others delve into case studies of individual states.

In Chapter 2, Fiona Davidson and Kimberly Johnson Maier examine the crucial role of women in the election, which was held exactly a century after women first gained the right to vote in 1920. Without the women's vote, Biden would have lost. They situate women's votes within the long history of the suffragist movement, which culminated in the election of the country's first female vice president. Importantly, they describe not simply how women voted, but also discuss their roles in mobilizing voters at the local and state scales, increasing turnout and catalyzing political participation.

The third chapter, by Jakob Hanschu and Laurie Johnson, focuses on the radicalization of right-wing voters in light of neoliberal globalization. American labor markets have been thoroughly restructured, with a decline in secure, middle-class jobs and the rise of the "precariat" of part-time and temporary workers. Economic insecurity has been accompanied by a steady stream of racist and anti-immigrant diatribes, all of which fulfil deeply seated needs for identity and feelings of belonging.

Chapter 4, authored by Adam Dohrenwend, addresses the voting gap between presidential and congressional candidates. The polls that confidently predicted a Democratic wave were misplaced, and Republicans did far better than expected. Although Trump added 10 million votes, down-ballot Republicans did even better, attracting ticket-splitters. Using a series of case studies, he illustrates that support for many Democrats was tepid; indeed, they lost more than a dozen seats in the House. Voter turnout among Democrats declined, especially among minorities. The distribution of voter drop-off and ticket-splitting, however, was unevenly distributed across the country.

The Senate elections in Georgia were pivotal to giving Democrats unified control of government. In Chapter 5, Fred Shelley examines the dynamics of the election in this state in detail. He notes the growing importance of the Atlanta metropolitan area in changing the politics of what was once a deeply red, but is now an increasingly purple, state. Biden was the first Democratic candidate to capture Georgia since 1992. Moreover, Georgia's voters ousted two incumbent Republican Senators and replaced them with Democrats Raphael Warnock and Jon Ossoff.

Another case study of a state, this time Wisconsin, is presented in Chapter 6, by Ryan Weichelt. In 2016, Wisconsin's defection to Trump shocked the political commentariat. Four years later, Biden narrowly recaptured the state, by just 20,000 votes. The state that elected Scott Walker and Ron Johnson had flipped back to the Democrats. The chapter examines this event in light of the Covid pandemic and the growing urban–rural schism in the state, which shaped the gubernatorial and primary elections as well as the presidential one.

Kenneth French and Ryan Weichelt argue in Chapter 7 that another instance of police violence against Black people, the shooting of Jacob Blake in the small working-class city Kenosha, Wisconsin, had repercussions that reverberated across the state and even had national implications. Riots protesting the shooting were condemned by Trump as domestic terrorism. Shortly thereafter, 17-year-old Kyle Rittenhouse attended the protest and shot three people, killing two. His subsequent arrest and trial received enormous national attention. In the 2020 election, places most affected by the riots voted most heavily for Trump. The case illustrates how local events and national politics can become deeply entangled with one another.

One of the unique features of the 2020 election was that it took place in the midst of the Covid pandemic. For many voters, it was the most important issue, and Trump's mishandling of the pandemic contributed significantly to his electoral loss. Shaun Johnson, in Chapter 8, summarizes the impacts of the pandemic, which varied among places. Early and mail-in voting soared. By comparing 2016 and 2020 voting patterns, he notes how small changes due to the pandemic had large implications in a close election. While Covid did not reduce turnout in rural areas that supported Trump, the pandemic alienated many suburban voters who opposed him.

For Canadians, observing the U.S. election was, as Jamie Essex puts it in Chapter 9, a bit like living in an apartment above a meth lab. Many Canadians viewed the turmoil of the Trump administration with alarm. Essex examines how roughly 620,000 American expatriates living in Canada (of whom he is one) participated in the election, how border restrictions brought on by the pandemic affect trans-border linkages, and how the Trump campaign reshaped political messaging in Canada. The result is an illuminating glimpse into a transnational political space.

In Chapter 10, John Paul Henry and Abraham Stephenson offer another trans-national perspective on the election, this time one based in Cuba. They discuss the role of Cuba in American politics, which reflects the long history of tense relations between the two countries. In 2020, rhetoric about Cuba reached new heights, with allegations that Democrats were socialists. Cuban-Americans have long been a potent and conservative political force in Florida politics whose votes have national repercussions. Their chapter delves into the affective dimensions of this issue, including human rights activists in both nations.

Chapter 11, by David Beard and John Heppen, compare Trump's strategy to that used in professional wrestling. Trump cut his political teeth in this world and adopted much of its showmanship. Thus, debates with Biden were "promo" events, his rallies were designed to "build the heat" or generate excitement, and claims of voter fraud led to the expected "rematch" in 2024 after he has left the ring.

References

Agnew, J. 1995. The rhetoric of regionalism: The Northern League in Italian politics. *Transactions of the Institute of British Geographers* 20:156–172.

Agnew, J. 1996. Mapping politics: How context counts in electoral geography. *Political Geography* 15:129–146.

Archer, J. 1988. Macrogeographical versus microgeographical cleavages in American presidential elections: 1940–1984. *Political Geography Quarterly* 7:111–126.

Archer, J., G. Murauskas, and F. Shelley. 1985. Counties, states, sections, and parties in the 1984 presidential election. *Professional Geographer* 37:279–287.

Archer, J. and F. Shelley. 1988. The geography of U.S. presidential elections. *Scientific American* 259:44–53.

Baccini, L., A. Brodeur, and S. Weymouth. 2021. The COVID-19 pandemic and the 2020 US presidential election. *Journal of Population Economics* 34(2):739–767.

Bekaroğlu, E. and G. Osmanbaşoğlu. 2021. Introduction to Turkey's electoral geography: An overview since 1950. In E. Bekaroğlu and G. Osmanbasoğlu (eds.) *Turkey's Electoral Geography*, pp.1–23. London: Routledge.

Eggers, A., H. Garro, and J. Grimmer. 2021. No evidence for systematic voter fraud: A guide to statistical claims about the 2020 election. *Proceedings of the National Academy of Sciences* 118:45–52. https://www.pnas.org/content/pnas/118/45/e2103619118.full.pdf

Forest, B. 2018. Electoral geography: From mapping votes to representing power. *Geography Compass* 12(1):12352.

Johnston, R. 2005. Anglo-American electoral geography: Same roots and same goals, but different means and ends? *Professional Geographer* 57:580–587.

Johnston, R. and C. Pattie. 2006. *Putting Voters in Their Place: Geography and Elections in Great Britain*. New York: Oxford University Press.

Johnston, R., F. Shelley, and P. Taylor (eds.) 1990. *Developments in Electoral Geography*. London and New York: Routledge.

Levitsky, S. and D. Ziblatt. 2018. *How Democracies Die*. New York: Penguin House.

Martis, K. 1982. *The Historical Atlas of United States Congressional Districts: 1789–1983*. New York: The Free Press.

Martis, K. 1989. *The Historical Atlas of Political Parties in the United States Congress: 1789–1989*. New York: Macmillan.

Martis, K., F. Shelley, J. Clarke, and G. Webster (eds.) 2014. *Atlas of the 2012 Elections*. Lanham, MD: Rowman and Littlefield.

Morrill, R. 1981. *Political Redistricting and Geographic Theory*. Washington, DC: Association of American Geographers.

Paulson, A. 2021. Stability and change in the 2020 election. In W. Crotty, A. Benner, and J. Berg (eds.) *The Presidential Election of 2020: Donald Trump and the Crisis of Democracy*, pp. 105–129. Lanham, MD: Lexington Books.

Prescott, J. 1959. The function and methods of electoral geography. *Annals of the Association of American Geographers* 49(3):296–304.

Reményi, P., H. Gekić, A. Bidžan-Gekić, and D. Sümeghy. 2021. Electoral geography of Bosnia and Herzegovina–is there anything beyond the ethnic rule? *East European Politics* 37(4):1–27.

Shin, M. and J. Agnew. 2008. *Berlusconi's Italy: Mapping Contemporary Italian Politics*. Philadelphia: Temple University Press.

Swauger, J. 1980. Regionalism in the 1976 presidential election. *Geographical Review* 70:157–166.

Warf, B. 2009. The U.S. electoral college and spatial biases in voter power. *Annals of the Association of American Geographers* 99(1):184–204.

Warf, B. (ed.) 2021. *Political Landscapes of Donald Trump*. London: Routledge.

Warf, B. and J. Leib (eds.) 2011. *Revitalizing Electoral Geography*. Aldershot: Ashgate.

Wing, I and J. Walker. 2010. The geographic dimensions of electoral polarization in the 2004 U.S. presidential vote. In A. Paez, J. Gallo, R. Buliung, and S. Dall'erba (eds.) *Progress in Spatial Analysis: Methods and Applications*, pp. 253–284. Berlin: Springer-Verlag.

Zarycki, T. 2015. The electoral geography of Poland: Between stable spatial structures and their changing interpretations. *Erdkunde* 69(2):107–124.

2 From Voting to Vice President
100 Years of Women in U.S. Politics

Fiona M. Davidson and Kimberly Johnson Maier

I am deeply conscious of the responsibility resting upon me.

Jeannette Rankin 1916

When Jeannette Rankin won an at-large congressional seat in Montana on November 7, 1916, she became the first woman elected to national office in the U.S. and the first (and to date only) woman elected to Congress from Montana. Coming only two years after Montana became the 11th state to grant women the right to vote in national and state elections, the Republican pacifist, suffragist and birth-control activist was keenly aware that her actions in office would be scrutinized in ways that had little to do with her politics and everything to do with her gender. Within a few years, the 19th Amendment ensured that women across the US could both vote and run in national elections, but it was 100 years after Rankin before a woman was to run (unsuccessfully) for president, and 100 years after the 19th Amendment before a woman would come within a heartbeat of the presidency.

History of Women's Suffrage in the U.S.

In 1776, Abigail Adams wrote to John Adams "In the new code of laws I desire you would remember the ladies and be more generous and favorable to them than your ancestors." Through their writing of the Constitution, the Founding Fathers laid the groundwork for determining who was and was not considered a citizen and ultimately decided that White women and people of color would be barred from political participation through informal and formal customs and laws.

The myth of the 1848 Seneca Falls meeting of abolitionists has come to dominate social memories of the (White) women's suffrage movement in the U.S., but it is important to consider and incorporate others, including Black men and Black women, into the broader history of women's suffrage (Tetrault 2014). Prior to the Civil War, Black men and women started organizing through anti-slavery organizations, such as the American Anti-Slavery Society. Black activists began discussing a variety of issues related to civil rights, employment, and voting in civil rights conventions as early as the

DOI: 10.4324/9781003260837-2

1830s (Dudden 2011), but it was not until the 1840s that the discussion of Black women suffrage was considered. In the 1850s, it became clear that Black women's suffrage was not a concern at Black conventions as Black men were more concerned with the impacts of legislation, such as the Fugitive Slave Law.

In the post-Civil War era, questions over who would be enfranchised first, White women or Black men, remained in question. Indeed, in some instances the two movements worked together; for example, Susan B. Anthony and Elizabeth Cady Stanton worked alongside Black men to achieve universal suffrage (Dudden 2011). However, the events following the Civil War created a clear division between abolitionists and women's suffrage groups.

Following the Civil War, the American Equal Rights Association (AERA) was formed, providing the first effort to organize suffrage for all (Black men and all women) at the national scale. The efforts put forth by this group were set back after an 1867 election in Kansas where Black suffrage and women suffrage appeared on the same ballot, ending in both being defeated. Consequently, there was a shift in the women's suffrage movement where Stanton and Anthony, who began working with racist Democrats, abandoned the idea of universal suffrage to focus on (White) women suffrage (Dudden 2011).

For most of the women's suffrage movement Black women existed in between the groups that were focused on Black male suffrage and those that focused on White women's suffrage (Dudden 2011; Tetrault 2014). Dudden (2011, p. 19) writes "Black women could fight for gender equality at black conventions or for race equality at women's rights conventions." Notable Black women such as Sojourner Truth and Maria Stewart spoke out for both Black and women's rights (Jones 2020; Tetrault 2014). For example, Sojourner Truth challenged White women over their racism and Black men over their sexism (Dudden 2011) and in the end, Black women created grassroots suffrage organizations through charitable and church organizations (Jones 2020).

With the passage of key Constitutional amendments, including the 14th Amendment (1868), which extended Constitutional protection to all "men," and the 15th Amendment (1870), which gave suffrage to Black men, further divided the suffrage movement. Key leaders in the women's suffrage movement, such as Stanton, began using racist rhetoric pertaining to Black-on-White rape, which resulted in continuing division within the AERA. Consequently, the women's suffrage group split in two. Susan B. Anthony and Elizabeth Cady Stanton created the National Woman Suffrage Association with the goal of advocating for a universal suffrage amendment in the U.S. Constitution as they opposed the 15th Amendment. The American Woman Suffrage Association, a pro-15th Amendment group, led by Lucy Stone and Henry Blackwell, on the other hand, took a state-by-state approach. In 1890, the two groups rejoined their efforts under the National American Woman Suffrage Association.

Over the course of the suffrage movement, White women organized, paraded in the streets, picketed, and participated in hunger strikes.

For example, on March 3, 1913, the day before President Wilson's inauguration, women organized and marched the inauguration parade route to show their exclusion from the democratic process. While Black women were present at this event, they were relegated to the back of the line to placate Southern sympathizers (Dudden 2014). Ultimately, they were left on the periphery because White women in the South needed to ensure voters that women's suffrage would not result in the enfranchisement of Black women.

While women's suffrage passed in 1920, it did not eliminate state and local voting restrictions that impacted women of color's access to the vote. For example, literacy tests plagued Blacks seeking the vote in the South. It was not until the passage of the Civil Rights Act (1964) and the Voting Rights Act (1965) some 45 years after the passage of the 19th Amendment that Black women were able to vote. Additionally, other ethnic groups, such as Asians, Latina, and Indigenous women, also encountered significant resistance accessing the ballot box. For example, Indigenous people were not given the right to vote until 1962 when all states extended suffrage to the Native American population. Additionally, all Asian Americans were not granted suffrage until 1952, when the Naturalization Act was repealed, and Hispanic Americans were not able to fully vote until 1975, when the Voting Rights Act was extended to include citizens with linguistic barriers. These expansions of the franchise contrast with more recent voting rights activity which includes the 2013 *Shelby County v. Holder* Supreme Court decision, which removed key elements of the of the Voting Rights Act, including the federal review process of closing polling places, limiting Voting IDs, and providing proof of citizenship. The modifications of state voting laws that are now permitted disproportionally impacted people of color, including Black women. Despite these limitations, Black women continue to have an impact on organizing voting efforts, which was particularly apparent in the 2020 presidential election.

Women as Voters

In 1920, 26.8 million American voters went to the polls to elect Warren G. Harding. This was by far the largest popular vote in U.S history, but it simultaneously represented one of the lowest voter turnouts, as the percentage of eligible voters who participated in the election plummeted from 61.6% in 1916, to just over 49% in 1920. This dichotomy, of course, represents the impact of the 19th Amendment on the U.S. voter landscape. The number of voters increased, but because the proportion of women who voted was far smaller than the proportion of men that voted; overall turnout dropped dramatically, and 1920 represents one of the three lowest turnout years in US voting history (the others being 1924 and 1996).

There are no national data on turnout by gender until exit polling began in 1952 (Gallup 2017), but small-scale local studies have indicated that depressed turnout was a function largely of women either fearing repercussions (both domestic and work-related) if they voted or, alternatively, holding anti-suffragist positions (Allen 2009). By the early 1960s, women's participation in presidential elections had increased to over 67%, but this was still lower than the turnout rate for men (71.9%). For the next three election cycles, turnout fell for both groups, but it fell much faster for men than for women, and finally, in the 1980 election, women outvoted men by a scant 0.3%, (59.4% to 59.1%) (CAWP 2017). Confoundingly, the 1964 election is often heralded as the first election in which more women voted than men, and while this is true, it is of course a function of the fact that there are consistently more voting-aged women in the US than men (currently women outnumber men in the U.S. by 7.3 million).

However, the tendency for women to vote at higher rates than men has been a constant in every election since 1980 and for the last nine presidential elections women have consistently outvoted men by an increasing margin. In 2020, 82.2 million women voted in the presidential election, the largest number of women who have ever voted, making up 52% of the electorate and accounting for 10.2 million more votes than were cast by men. These women represent 68.4% of all eligible female voters and female turnout increased from 2012 (63.7%) and 2016 (63.3%) and was 3.4% higher than turnout for men, a gap that narrowed slightly over recent elections (3.9% in 2012 and 4.0% in 2016) (CAWP 2021b).

The combination of both more women in the population and a tendency for them to vote in higher proportions than men has resulted in increased focus on the "gender gap," which highlights the different partisan leanings of men and women in the U.S. population. This difference in male and female partisan voting is first captured by exit polls in 1952 (Gallup Historical polls), when 5% more women than men voted Republican. This Republican bias continued through the next two election cycles until 1964 when women favored Johnson over Goldwater by 24 points, and for the first time had a higher Democratic vote than men by 2 percentage points. In every election since 1964 – except for the Carter/Ford election of 1976 and the 1992 Clinton/Bush election, when substantially more men voted third party than women – women have been more likely to vote Democratic than men, with margins ranging from 1% (1972) to 12% (2016 and 2020). While there have been fluctuations in the size of the gap over time, overall, it has increased steadily, averaging around 4% in the 1970s, and 12% over the last two elections. In 2016, 54% of women voted for Hillary Clinton, an increase of 1% over the female vote for Obama in 2012, and the gender gap reached an historic 12%, or 24 points (the sum of the difference between the female/male Republican and Democratic votes) (Archer et al. 2014; Watrel et al. 2018).

Women Voters in the 2020 Election

In 2020, this partiality continued, with 57% of women voting for Biden, as opposed to 45% of men, again representing a 12% gap between male and female preference for the Democratic candidate. The relative lack of third party voting in the 2020 election resulted in a Republican gap of 11% (42% of women voted for Trump versus, 53% of men).

There is relatively little state-scale difference in the total votes for Biden and the female vote for Biden. In every state except for West Virginia, women were more likely to vote Democratic than men, but the overall pattern mirrors the national vote, with women less likely to vote Democratic in strongly Republican states (the South and Midwest) and more likely to vote Democratic in strongly Democratic states (the coasts and Hawai'i).

At the state level, the partisan preferences of male and female voters can be compared by looking at the male/female partisan split. In 2016, there are 19 states with a split vote (men voting majority Republican and women voting majority Democratic), but by 2020 that was reduced to 16 as there occurred a consolidation of Republican states from Idaho to West Virginia. Where in 2016 several of these states demonstrated a split in the vote (Republican win, but with Democratic majority from women voters), the 2020 vote shows Republican majorities from both men and women voters. Several other states, notably Colorado and Virginia, experienced a shift from a split vote (men Republican/women Democratic) to both genders voting Democratic, but most important are the states that shifted from Republican win/women Democratic to Democratic win/women Democratic. In all these states the gender split did not change – a majority of men voted Republican, and a majority of women voted Democratic, but the relative balance between the two changed sufficiently that Wisconsin, Michigan, Pennsylvania, and Georgia shifted from Trump in 2016 to Biden in 2020. This change does not mean, however, that women were responsible for flipping those states, in Pennsylvania, the female Democratic vote remained static from 2016 to 2020 (at 55%), it was the increase in the male Democratic vote, from 40 to 44%, that flipped the state, and in Georgia, both men and women were 6% more likely to vote for Biden than for Hillary Clinton four years earlier (Archer et al. 2014; Watrel et al. 2018).

The overall similarity between the total vote and the female vote begins to break down when women are disaggregated by age and ethnicity. Women are by no means a monolithic voting block and exit polls indicate that White women were not only more likely to vote for Trump than for Biden (55% to 44%), but that the Republican preference increased by 1% from the 2016 election. Across the country, in only nine states did White women vote at, or above, the national average for women voting Democratic; in all other states White women underperformed women in general as Democratic voters. This Republican partisanship among White women demonstrates a clear regional pattern (Figure 2.1). States in which White women were much less likely to vote Democratic than women of color (in the case of Mississippi by a stunning 78%) are clustered in the South, from Arkansas and Mississippi to South Carolina and West Virginia; the closest any Southern state gets to the national average for the female Democratic vote (57%) is Florida, where 40% of White women voted for Biden. In those Southern states, the only one to break for Biden was Georgia, where only 32% of White women voted for Biden, 60 points less than the percentage of women of color who voted Democratic, demonstrating once again the tremendous importance of the Black female vote in swinging Georgia blue.

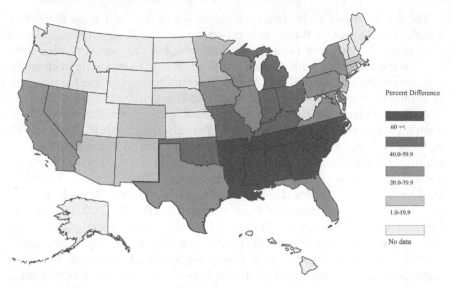

Figure 2.1 Gap Between White Female Democratic Vote and Women of Color Democratic Vote in 2020.

Source: authors.

The gap between the Democratic vote for White women and women of color is also substantial in a swath of Republican states that surround the South, from Nebraska to North Carolina, and still considerable in a third group of states that stretches from Pennsylvania to Arizona. This group is particularly interesting because it includes three of the five battleground states in the 2020 election, two of which were successfully flipped for the Democrats. In both Pennsylvania and Arizona, the relatively low White female Democratic vote (47% and 46% respectively) was countered by the votes of women of color who turned out in metropolitan areas and native tribal lands in record numbers, voting 65% for Biden in Arizona and 91% for Biden in Pennsylvania.

Finally, it is worth noting that there is only one Republican state in which White women vote close to the national average, and that is Alaska (53%); all of the other 19 states where at least half the White women vote Democratic were won by Biden.

The explanation of this regional gap in the way women vote is complex but relates to both voter location and identity. In states with largely rural and small metropolitan areas (Alabama, Arkansas, West Virginia, Mississippi), the underlying White/Republican Black/Democratic voting dynamic is long-standing and well-documented (Maxwell 2015) and translates into states where the predominantly White rural/small town/suburban population produces women voters who are much more likely to be religious, traditional, married and older, all factors that increase Republican partisanship. With relatively small urban populations there are not enough educated, single, non-religious and younger women to raise the level of the White female Democratic vote closer to that of the average female vote, which is bolstered by the more reliable Democratic leanings of women of color. At the same time, even for younger urban voters in the South, there is evidence that the tendency for White women in the South to prioritize race over sex in their self-identification reduces the White female Democratic vote throughout the region (Maxwell and Shields 2019). Consequently, in states like Mississippi and Alabama, the gap between White and Black women's votes for Biden was 79 points and 74 points, respectively (Mississippi, 16% White and 95% Black; Alabama, 19% White and 93% Black). Outside the South the gap narrows either because there are so few women of color, that their influence of the average female vote is not significant (the Great Plains, Rocky Mountain and New England states) or because the presence of major metropolitan areas with their populations of young, single educated White women, increases the White female Democratic vote to approach (although never equal) the Black female Democratic vote.

Black Women Voters in the 2020 Election

It is important to remember that despite the consistently high levels of support for Democrats, Black voters are no more monolithic than any other group. While there is relatively little geographic variation in the Black vote – all counties with large Black populations demonstrate support for the Democrats – within the Black voting population there is one particularly important cleavage, the difference between male and female voting patterns. In the 2020 election, Black men were less likely to vote for Joe Biden than were Black women by 14 points (79% to 93%), although it is important to note that Black women were twice as likely to vote as Black men (making up 8% of the total vote, while Black men accounted for only 4%).

Much has been made of the increased Black vote for Trump, which increased by 3% for men and 4% for women between 2016 and 2020, but except for the record Black vote for the Democratic Party in the Obama elections (2008 and 2012), these current numbers are entirely in line with Black voting over the last 40 years. Since the early 1970s Black men have historically been much more likely to vote Republican than have Black women; in 1972, 23% of Black men supported Nixon (Wright Rigeur 2020a). Support for the Republicans among Black men declined during the Reagan/Bush presidencies but rebounded again under Clinton and the second George Bush. Black women, on the other hand, are much more reliably Democratic, with their support for Democratic candidates never falling below 90% in the past half century.

The reasons for this gendered cleavage seem to be rooted in the different ways in which Black men and women perceive community interest, with Black men marginally more likely to buy into the Republican ethos of voting in pursuit of policies that produce individual gains, while Black women are more likely to vote Democratic in support of policies that benefit the entire community (Wright Rigeur 2020a).

While there is relatively little spatial variation in the tendency for Black women to vote Democratic, there is, of course, considerable spatial variation in the location of Black women available to vote. The concentration of the Black population in the United States is largely a function of two historical contexts: slavery and Black rural to urban migration during the 20th century. These circumstances have created one of the most distinctive geographical patterns of any ethnic group in the U.S. with the Black population concentrated in a band of rural counties across the South from the Mississippi Delta through the Black Earth belt of Alabama and Georgia into the urban tidewater counties of the Atlantic seaboard and Northeast. With the exception of Wayne County, Michigan (Detroit), counties where the Black population makes up at least a third of the total population form a crescent from northern Louisiana to southeast Pennsylvania, passing through most of the major metropolitan areas of the South and East (Birmingham, Memphis, Atlanta, Charlotte, Washington DC, Philadelphia). It is in these metropolitan areas, with their motivated, determined Black voters, that the 2020 election was decided, and where much of the succeeding controversy about voter fraud was concentrated.

While almost all counties in the U.S. trended Democratic in the 2020 election, it was the scale of turnout and partisan support for the Democrats in Milwaukee, Wayne, Dekalb, Fulton, Clayton, Philadelphia and Montgomery counties with their large populations and overwhelming support for Joe Biden that clinched Democratic victories in four critical states and delivered the 72 Electoral College votes that ensured Joe Biden's win.

The importance of the Black vote, especially the activism of Black women, in the 2020 election, cannot be overstated. In cities across the U.S., especially in Philadelphia, Detroit, Milwaukee and Atlanta, the movement to register and mobilize the Black population was critical to the narrow margin of Joe Biden's wins in Pennsylvania, Michigan, Wisconsin and Georgia. The mobilization of the Black population in Georgia's runoff elections, created the narrow Senate victories for Democrats Warnock and Ossoff in January 2021.

Finally, it is not possible to discuss Black women in the 2020 election without addressing the groundbreaking candidacy of Kamala Harris. A successful California state prosecutor, Harris became the first African American to win a California Senate seat in the 2016 election and parlayed that success into a run for the Democratic nomination in early 2019. While her support rose after the first Democratic presidential debate in June 2019, a focus on her tenure as California Attorney General resulted in criticism and falling poll numbers in the latter part of that year and she dropped out of the race in December 2019, throwing her support behind Joe Biden. Discussed as a possible running mate for Biden as early as May 2019, the Biden-Harris ticket was announced in summer 2020, with Biden taking advice from several members of the Congressional Black Caucus that it was time to reward Black women for their consistent support of the Democrats (Timm and Gregorian 2020). Criticized for her inability to mobilize the Black community to back her presidential candidacy, Harris's support suffered both from her reputation as a state prosecutor, and from a pragmatic fear among many Black voters, especially Black women, that the average American voter was not yet ready to support a Black female for President (King 2019; Wright Rigeur 2020b). At the same time, Joe Biden's links with President Obama ensured that he did have the support of the Black community, and when linked to a female Black vice-presidential candidate this ticket virtually guaranteed the enthusiastic mobilization of Black women that resulted in the narrow successes in the urban centers of swing states that were so critical to the Democratic victory.

Other Women of Color in 2020

Black women are more reliably Democratic than any other demographic; however, all non-White demographic groups, both male and female, share a partisan preference for the Democrats. Historically, Latinx, Asian and Native American, as well as those who identify as mixed race, have voted consistently for Democratic candidates at both the national and local scales.

In 2020 for the first time Latinx voters became the largest minority voting bloc in the US (13.3% of all voters compared to 12.5% for Black voters). While the group has a whole consistently demonstrates the lowest turnout rates in presidential elections, turnout increased by 6% in 2020 (to 53%) and by as much as 18% in some critical battleground states (Michigan and Arizona) (Brookings Institute, 2021). At the same time, female Latinx voters continued to outvote their male counterparts by a further 5%, so while the 2020 Latinx vote remained consistent with the vote in 2016 (65% Democratic, 35% Republican), the gender gap in the Democratic vote increased from 6% (63% male, 69% female) to 10% (59% male, 69% female). Consequently, while the most discussed aspect of Latinx voting in the 2020 election was the significant rise in the Republican vote of heavily Latinx counties in southern Florida and along the Texas, Mexico border, the overall impact of Latinx voters, especially given increased turnout in the Midwest and West, was to bolster the Democratic vote in battleground states like Michigan and Arizona and contribute to the Democratic victory (UCLA Latino Policy and Politics Initiative 2021).

The contributions of Asian and Native American women are harder to gauge, given the limited number of states in which their presence rises to a level that provides useful exit poll data. The Asian population as a whole trends Democratic, with a consistent (since 1980) 65% of the Asian voters backing Democrats; however, there is no observed tendency for women to outvote men and no appreciable partisan gender gap (Xiao and Bass 2021). Similarly, the Native American population, despite its reliable tendency to vote Democratic, tends to be statistically insignificant in the face of the overwhelming White, rural, Republican voters of the Dakotas, Arizona and Alaska. Despite this, there is evidence in the 2020 election that greatly increased Native American turnout (up as much as 116% over 2016) as well as concerted efforts to overcome voter suppression efforts resulted in small

Democratic gains that contributed to overall victories in states like Arizona and Wisconsin, with some tribal precincts in the Tohono O'odham Nation of Arizona breaking 98% for Biden/Harris (Smith 2020).

Overall, the 2020 election saw a continuation of several longstanding voting patterns, with women generally preferring Democratic candidates, while reminding us that women are not a monolithic voting bloc. As in 2016 and previous elections, there was significant spatial variation in the way that women voted and considerable divergence between the way White women and women of color voted.

Women as Candidates

While women have been active as voters since before the 19th Amendment (many states permitted some, if not all, women to vote in local and state elections prior to 1920), widespread female participation as candidates is a much more recent phenomenon in the U.S. At the national level, the first woman sworn into the House of Representatives was Jeannette Rankin (R) from Montana in 1917 and in the Senate, 87-year-old Rebecca Felton (D) of Georgia, who was appointed to fill a vacancy and served just 24 hours in 1922. The pattern of women being appointed to fill positions vacated by their husbands and fathers continued throughout the 20th century and includes current Alaska Senator Lisa Murkowski (R), who was appointed to fill her father's seat when he resigned in 2002, although she has since won three elections in her own right.

Since 1992 the Center for Women in American Politics (CAWP) at Rutgers University has been collecting data on female candidates for national and state office in the U.S., creating an invaluable database of women candidates over the last 30 years. Covering the results of both primary and general elections, at the state and national scale and including partisan affiliation, this database provides the foundation for the sections on women as candidates in 2020.

There was a great deal of press attention when the 2020 election represented a high point for women's candidacy for national office, with a record 583 women running for the House and 60 for the Senate (these records include both primary and general election numbers). However, 50% of these women lost their primaries, and over 60% lost their general election runs. Overall, women are both less likely to run for national office (only 35% of the candidates for House and Senate elections in 2020 were women) and even less likely to win (27% of winning candidates were women) when they do run (CAWP 2021a). Consequently, an examination of the current House and Senate indicates that, as of January 2021, women made up only 26% of the combined membership. At the state level, there are nine states with no female representation (one of which, Vermont, has never had a woman elected to national office) and only five states (Iowa, New Hampshire, Nevada, Washington, and Wyoming) with more women than men in their delegations. There is no clear partisan pattern to this representation, while Democratic states like California and Washington have higher percentages of women in their delegations, so do Republican states with small delegations and popular, long-serving female candidates (Alaska, Iowa, and Wyoming). However, in general, the larger the state population, and the more urban the state, the more likely that the Congressional delegation will include a substantial number of women. California holds the record for female representation.

Overall, female representatives are, not surprisingly, given the policy positions of the parties and their increasingly gendered voter support, much more likely to be elected as Democrats than Republicans. Currently, there are almost twice as many women in the Democratic Caucus (39%) as in the Republican (27%), and 73% of all the women in Congress are Democrats. Despite the record number of Republican women elected in 2020, the imbalance between Democrats and Republicans is becoming more pronounced over time as any increase in Republican women elected is dwarfed by greater increases in Democratic women. This has not always been the case; both the first woman elected to the House and the Senate were Republicans, and in the history of women's representation in both bodies, Republicans accounted for 35% of all the women elected. However, in recent years the increased mobilization of Democratic women has outpaced the ability of Republicans to nominate and elect Republican women.

Women also made records at the state level, after rapid increases in female representation in the 1980s and 1990s, which plateaued in the early years of the 21st century (Carroll 2013), women again began running and winning state legislative seats in the 2016, 2018 and 2020 election cycles with 3,583 female candidates in the 2020 state legislative elections, representing 24% of the total. Of those women, 1,191 won their races, and currently 28.9% of state legislative positions are held by women. Across the U.S. northern and western states demonstrate the highest representation for women in state legislatures and Southern states the lowest (Figure 2.2). These states demonstrate a clearer partisan pattern than the national delegation. Seven states, six of which are in the South or Appalachia, have less than 10% of their state legislative seats and they are dominated by Republican voters, whereas states with substantial female representation in their state legislatures (40%+) all voted Democratic in the 2020 Presidential election.

Women of Color as Candidates

Unsurprisingly, among all the other records that were set by women in both the 2018 and 2020 elections, the 2020 election saw a record number of women of color running for both state and national offices and saw a record number of them succeed in being elected.

It is important to note that this kind of electoral success is relatively new. The first woman of color elected to the House was Patsy Takemoto Mink, a Democrat from Hawaii in 1964, and Shirley Chisolm (D) of New York was the first woman of color elected to the House (in 1968). The current 74 women of color in the House and Senate comprise a majority (57%) of all the women of color ever elected to Congress.

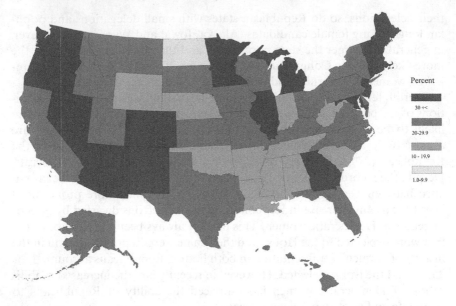

Figure 2.2 Women as Percent of State Legislatures, 2021.
Source: Authors.

The 2020 election resulted in the election of a record 124 members of underrepresented groups being elected to the 117th Congress, of whom 74 are women of color. These women represent 39.5% of the underrepresented population in the House and Senate and comprise 34% of all the women in Congress. While 34% is still considerably less than the 39.6% of the total population that is classified as being Latinx, Black or AAPI, female representation in the underrepresented groups, particularly in Black women (46% of the Black delegation) is considerably more robust than the 19% of White representatives who are women.

Geographically, Congressional representatives who are women of color are most likely to come from California, Washington, Nevada, and Hawaii, with a second concentration in the Northeast (New York and Delaware) and two mid-continent outliers in Minnesota and Kansas. A middle group of states in the South and Midwest shows the importance of representation of statewide minority populations, and populations that are concentrated in urban areas (Illinois and Michigan). Crucially, however, most of the states with a higher-than-average percentage of their delegation made up of women of color (12 of the 16) broke for Biden in the 2020 election. Of the four states that voted Republican, two are battleground states where significant numbers of Democrats elected four women of color in Texas and three in Florida, and the other two are solidly Republican states where, again, Democrats elected a single Native American Representative from Kansas, and a single Black woman Representative from Alabama. Across the U.S., the states that voted Democratic produced 62 female minority representatives while the

Republican states elected just 12, and of those 74 women of color, only four were elected as Republicans.

This partisan imbalance is a critical part of the geographical unevenness of the electability of women of color, and all marginalized groups. In Congress, people of color make up 43% of the Democratic caucus, while making up only 8% of the Republican caucus. This relatively high representation of marginalized groups in the party that is (albeit only by a few members) in the majority in Congress results in higher levels of success for women of color, who represent broad constituencies of diverse, usually urban, voters, relative to White women, whose prospects are more likely to be constrained by the more traditional, rural and often religious voters who vote Republican.

Women of Color in State Legislatures

While all marginalized groups continue to be significantly underrepresented in state legislatures, data from the 2020 election indicate that new records were set for women of color running in elections and being elected. In 1971, women comprised 4.5% of all statewide elected representatives and women of color less than 1%. Those numbers increased throughout the 1970s and 1980s only to plateau in the 1990s (Carroll 2013). It is not until the 2010s that the percentage of women, and women of color, began to increase again, with significant jumps taking place at the 2016 and 2018 elections (CAWP 2021b).

After the 2020 elections, there were 607 women of color serving in state legislatures nationwide, comprising 26% of all the women in state legislatures and 8.2% of all state legislators. Every state has at least one woman of color in the state legislature and Hawai'i has the largest proportion of women of color, with 41% of their state legislative seats held by women of color. The pattern of representation by women of color at the state level (Figure 2.3) is broadly similar to that of women of color in state delegations to Congress. States with the highest levels of representation have larger minority populations; however, Democratic states with relatively low minority populations (Washington, Colorado, Nevada) also show higher than average percentages of women of color in state legislatures. Conversely, states with small minority populations have much smaller minority women representation. However, if we look at representation relative to the state's minority women population a more interesting pattern emerges.

An analysis of legislative representation relative to total population shows that very few states have an equal or higher proportion of women of color in state legislature than women of color in the overall population, only ten states fall into this category, Colorado and Nevada have the largest gaps with 5% more women of color in their legislatures than in the general population. Of the ten, only Florida and Texas broke for Trump in the 2020 election, both states had relatively robust percentages of their Latinx population who voted Republican and both states have Latinx women in the state Republican Caucus. At the other end of the spectrum, all three of the states that demonstrated huge (more than 10%) gaps between the percentage of women of

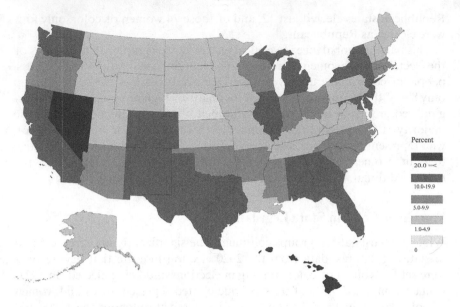

Figure 2.3 Women of Color as Percent of State Legislatures, 2020.
Source: Authors.

color in the population and in the state legislature are solidly Republican states, three in the South/South Central US and Alaska. There are no strong patterns in the middle groups; geographically; they are scattered from California to Maine and they exhibit an almost even partisan split (16 Democratic/17 Republican).

Overall, in state legislatures, the partisan nature of representation for women and women of color means that states with Democratic majorities exhibit the highest levels of representation. In 2021, of the 2,290 women in state legislatures nationwide, 66% were Democrats and 34% were Republican, with the latter concentrated in Republican strongholds like Kansas, Idaho, Mississippi, and Texas. Wyoming is the only state where all the female legislators are Republicans, while there are no female Republicans in Delaware, Hawaii, Vermont, and Massachusetts. Women of color are even more likely to be Democrats; after the 2020 election, 95.6% of all state representatives that are women of color were elected as Democrats, with only 25 Republicans who are almost exclusively women who identify as Latinx or Asian.

Conclusion

It is clear that at both the national and state levels, as voters and as candidates, women generally vote, run and get elected as Democrats. It is unsurprising, then, that while the first woman elected to Congress was a Republican 105 years ago, the closest that any woman has come to the presidency (as unsuccessful presidential candidate in 2016 and as successful running mate in 2020) is always on the Democratic ticket.

The partisan nature of women's political preferences has changed slowly in the last 100 years and appears to be continuing to shift toward the Democrats. This is particularly true among women of color who are becoming more numerous in each electoral cycle and as a result of the long historical record of voter suppression against minority voters, more active, with predictable results for Democratic candidates, especially in tight races.

The activism and the narrow victories for Democrats, particularly in Georgia, have instigated another round of voter suppression measures in state legislatures around the country. Since January 2021, Republican state legislators have introduced over 400 different measures to restrict early days, limit mail-in voting, reduce ballot drops, even to criminalize the provision of food and water to people waiting in voter lines. While these measures are technically non-partisan and race-blind, they will differentially impact populations who are urban, poor, marginalized and minority, and, above all, who vote Democratic. As Alice O'Lenick, the Republican Chair of the Gwinnett County Board of Registrations and Elections, said in January 2021, in reference to voting laws in the state of Georgia "They don't have to change all of them, but they've got to change the major parts of them so that we at least have a shot at winning" (Wines 2021).

References

Allen. J. 2009. Reluctant suffragettes: When women questioned their right to vote. http://www.pewresearch.org/2009/03/18/reluctant-suffragettes-when-women-questioned-their-right-to-vote/

Archer, J., R. Watrel, F. Davidson, E. Fouberg, K. Martis, R. Morrill, F. Shelley, and G. Webster (eds.) 2014. *Atlas of the 2012 Elections*. Lanham, MD: Rowman and Littlefield.

Carroll, S. 2013. *The Book of the States 2013*. Lexington, KY: Council of State Governments.

Center for American Women in Politics (CAWP). 2017. Gender differences in voter turnout. http://www.cawp.rutgers.edu/sites/default/files/resources/genderdiff.pdf

Center for American Women in Politics (CAWP). 2021a. Black women in American politics 2021. https://cawp.rutgers.edu/sites/default/files/resources/black_women_in_politics_2021.pdf

Center for American Women in Politics (CAWP). 2021b. Elections database. https://cawp.rutgers.edu/facts/elections/past_candidates

Dudden, F. 2011. *Fights Chance: The Struggle over Women's Suffrage and Black Suffrage in Reconstruction America*. Oxford: Oxford University Press.

Gallup US Presidential Election Center. 2017. http://www.gallup.com/poll/154559/us-presidential-election-center.aspx

Jones, M. 2020. *Vanguard: How Black Women Broke Barriers, Won the Vote, and Insisted on Equality for All*. New York: Basic Books.

King, M. 2019. Why Black voters never flocked to Kamala Harris. *Politico* (December 4). https://www.politico.com/news/2019/12/04/kamala-harris-black-voters-2020-075651

Maxwell, A. 2015. Untangling the gender gap in symbolic racist attitudes among White Americans. *Politics, Groups, and Identities* 3(1):59–72.

Maxwell, A. and T. Shields. 2019. *The Long Southern Strategy: How Chasing White Voters in the South Changed American Politics*. Oxford: Oxford University Press.

Rankin, J. 1916. *United States House of Representatives: History, Art & Archives*. Retrieved April 12, 2018.

Smith, A. 2020. How Native American voters swung the 2020 presidential election. *Missoula Current* (November 13). https://missoulacurrent.com/government/2020/11/native-vote/

Tetrault, L. 2014. *The Myth of Seneca Falls: Memory and the Women's Suffrage Movement, 1848–1898*. Chapel Hill: University of North Carolina Press.

Timm, J. and D. Gregorian. 2020. Clyburn calls for democrats to 'shut this primary down' if Biden has a big night. *NBC News* (March 10). https://www.nbcnews.com/politics/2020-election/clyburn-calls-democrats-shut-primary-down-if-biden-has-big-n1155131

UCLA Latino Policy and Politics Initiative. 2021. Latino voters were decisive in 2020 presidential election. *UCLA*. https://newsroom.ucla.edu/releases/latino-vote-analysis-2020-presidential-election

Watrel, R., R. Weichelt, F. Davidson, J. Heppen, E. Fouberg, J. Archer, R. Morrill, F. Shelley, and K. Martis (eds.) 2018. *Atlas of the 2016 Elections*. Lanham, MD: Rowman and Littlefield.

Wines, M. 2021. After record turnout, Republicans are trying to make it harder to vote. *New York Times* (March 26). https://www.nytimes.com/2021/01/30/us/republicans-voting-georgia-arizona.html

Wright Rigeur, L. 2020a. *Black Voters, Gender, and the 2020 Election*. Atlanta: Black Women's Think Tank at Georgia Institute of Technology.

Wright Rigeur, L. 2020b. *Super Tuesday Results*. London: London School of Economics.

Xiao, H. and L. Bass. 2021. Who votes among Asian American ethnic subgroups. *Socius* 7:1–13.

3 The Economic and Psychological Origins of Right-Wing Radicalization in the U.S.

Jakob Hanschu and Laurie M. Johnson

The upswing in activity of alt-right, white nationalist, militia and related movements in the United States over the past decade took a dramatic turn with the insurrection at the Capitol on January 6, 2021. The brazen nature of the planning and communications that came before the attack, and the large assortment of right-wing groups and movements represented during the event, finally caused some commentators to consider the possibility that something more than racism, nationalism, or any other singular ideology, was at work (Aiginger 2020; Zakaria 2016). As more attention was focused on the process of radicalization through propaganda from sources like Q-Anon, Newsmax, Breitbart, TurningPoint USA, Fox News, the *Epoch Times*, Twitter, 4-Chan, Parler, etc., it became conceivable that we were looking at some sort of web-accelerated social disruption, something akin to what happens when people join cults, but at the mass level (Nagle 2017).[1] Most critics of these trends have spoken of them as though the individuals attracted to conspiracy theories and extremism were making willful choices to think and do bad things because of personal character flaws. Such critics fired back at MAGA-followers and adjacent groups with what they considered reason and facts, and not a little bit of anger and blame. To respond any other way would be to admit that the followers of such movements were not completely responsible for their radicalization, and this was not (or not *just*) a phenomenon of conscious, individual choice. It required looking into the *conditions* that produce these unrealistic, paranoid, and violent beliefs and behaviors, and re-imagining these followers as at least partly responding to those conditions. Tempting as it is to place the blame on individuals and move on, this chapter will refrain from condemnatory moralism in order to examine the roots of the protracted tragedy that is revealing itself in real time in the 2020s. The chapter concludes with some thoughts about how to deal with the mass psychosis involved in recent waves of right-wing radicalization.

Non-Stop Changes

The twin forces of automation and economic globalization have hit workers hard around the world. Negative effects of these forces include social and cultural destabilization, lower incomes, and more precarious modes of

DOI: 10.4324/9781003260837-3

sustenance (Appadurai 1996; Hardt and Negri 2000; Standing 2011). The increased suffering and anxiety subsequently arising from this general state of insecurity has led to various manifestations of mental health problems and sociopolitical unrest in different areas (Briggs 2021; Fisher 2008). We focus on the Midwestern region of the United States, where the manufacturing and agricultural sectors have undergone massive and rapid changes.

Since the Industrial Revolution, automation has served as a steady driver of the capitalist economy.[2] It promises increased efficiency and standardized quality, which means lower long-term costs of production for the owners of the technology and a profusion of new, always-cheaper goods for consumers. However, the effects of automation on income and job security have harmed many people while benefiting relatively few, mainly the owners and (less often) the creators of techno-logistical innovations and information (Marx 1976; Zuboff 2019). Makenzie Wark calls this new dominant owner class that specializes in the "information asymmetry business" the "vectoralist" class (2020, p. 45).[3] While the trends toward less manual labor and more flexible hours ushered in by automation might appear progressive, the negative effects have been decreasing job security, monotonous and even dehumanizing work, cheaper goods characterized by planned obsolescence, and growing consumer debt obtained by chasing the increasing amenities associated with the ideal middle-class lifestyle and the ever-receding horizon of financial security (Ehrenreich 2020; Konings 2018). Further, the same technological developments have led to the destabilization of the distinction between work and non-work activities and the emergence of an economy that is "always on," carrying with it the erosion of social life and a proliferation of anxiety and depression (Briggs 2021; Crary 2013).

While automation has tended to make work less fulfilling and more precarious, globalization has led to a massively increased labor pool and an equally rapid decline in real wages and opportunities (Piketty 2014). This problem is particularly more pronounced in the global North, where expectations for economic advancement continue to unrealistically increase (Beckert 2016). Young people in these countries sense that they have fewer opportunities than their parents, and probably will be less secure as they enter their retirement years (Aaronson and Mazumder 2008; Brown 2018; Chien and Morris 2018; Dettling and Hsu 2017). One reason for the relative insecurity of more recent generations is immense student debt. As Jeffrey Williams writes:

> Debt is not just a check every month but colors the day-to-day experience of my life, whether I live in a smaller or larger apartment, whether I can buy a house (not yet), whether I can travel to Europe (not since grad school), whether I can eat out (too often considering the debt).
>
> (2006, p. 156)

While Williams's statement is specific to student debt, it can be understood as applying to debt of all forms more broadly.

Globalization has increased the GDP of developing countries while decreasing their economic stability, particularly their populations' access to subsistence agriculture and social commons that previously allowed the money-poor a source of survival (Bensaïd 2021; van der Ploeg 2008).[4] Globalization has increased the GDP of developed countries even more dramatically while increasing the numbers of people with no job security or hope for a comfortable old age. Class inequality has increased in both developing and developed countries but has grown more rapidly in countries with emerging economies since 1990, making for increasing frustration and resentment from those in the West who cannot win in the current configuration of capitalism (Piketty 2014).[5] This trend has coincided with increased migrant flows into relatively wealthy countries, putting more pressure on cultural and community resources there.

This is where we are now – but how did this happen?

The labor movement in the US began to decline around the same time as small farms and small businesses were challenged by the US government's turn toward prioritizing corporate production and global markets (Aronowitz 2014). By the 1970s the expansion of the world economy in countries like Germany and Japan that had benefited from U.S. post-war aid and protection had begun to affect the domestic economy. US manufacturers were having a difficult time competing with foreign manufactured goods, leading to a drop-off in blue-collar jobs in the U.S. and reductions in wages, benefits and job security for those who remained. The Carter Administration ushered in a round of tax cuts and deregulation with the hopes of stopping this trend (Anderson 1991). Not only did this move fail; it also obscured the fact that the U.S. regulatory apparatus that remained favored large-scale producers over small and local producers of both raw and processed foods and goods (Hendrickson et al. 2019; Howard 2021). Lower corporate taxes and a friendly regulatory environment encouraged corporate growth, and cheap labor costs abroad encouraged corporations to locate manufacturing offshore (Harvey 2005).

In agriculture, the Nixon Administration made a conscious decision, articulated by Agricultural Secretary Earl Butz, to provide subsidies and supports to large-scale monoculture commodity crop farming (Paarlberg and Paarlberg 2000). It was not long before ethanol was mandated by many states as a prop for the giant corn farms that dominate the American Midwest, creating what Sautter et al. (2007) call "a fool's paradise" (see also Wallander et al. 2011). These events precipitated the demise of the diversified family farm, and small-town businesses and communities began to languish (Dimitri et al. 2005; Foltz et al. 2002; Gardner 2002; Lyson et al. 2001; Smithers et al. 2004). The tax-cutting policies of the Reagan Administration gave another huge boost to corporate business over small and middling proprietors, as did Supreme Court decisions such as *Kelo v. City of New London* (2005), which allowed local and state governments to use eminent domain law to clear out locally-owned businesses and provide tax abatements and free property enhancements to bring in corporate big-box stores. By the time Ronald

Reagan confronted the air traffic controllers' union and Republicans began to pass state "right to work" laws, the deck was already stacked against entrepreneurs and small startups, as well as workers who wanted stable well-paying jobs and the ability to organize. As the economic and political power of labor faded, corporate economic and political power continued to grow. The more political power corporations wielded, the more Republicans and Democrats alike aligned themselves with the need to reduce barriers to international trade to make the world "safe" for the globalized capitalist market, leading some critics to argue that the left has abandoned its fight for the working class in favor of culture war issues (Lasch 1996; Nakamura 2015).

When few working-class and lower- to mid-level white-collar jobs come with any real security, and when most large items like cars and homes often are never fully owned, one can argue that a large chunk of the American public falls into the category of the "precariat." The term brings together the adjective "precarious" with the noun "proletariat" to denote a class of people inhabiting a status position that lacks labor-related security and predictability (Standing 2011). As Anna Tsing writes: "Precarity once seemed the fate of the less fortunate. Now it seems that all our lives are precarious – even when, for the moment, our pockets are lined" (2015, p. 2). Life amid such indeterminacy often leads to deleterious effects on socioeconomic and psychological wellbeing (Standing 2011). The precariat is not homogenous, though. Standing (2011, pp. 13–14) writes:

> The teenager who flits in and out of the internet café while surviving on fleeting jobs is not the same as the migrant who uses his wits to survive, networking feverishly while worrying about the police. Neither is similar to the single mother fretting where the money for next week's food bill is coming from or the man in his 60s who takes casual jobs to help pay medical bills. But they all share a sense that their labor is instrumental (to live), opportunistic (taking what comes), and precarious (insecure).

Similar to Marx's (1976) "proletariat," members of the precariat lack property, must sell their labor to subsist, and experience high levels of exploitation. However, their diversity of occupations, lifestyles, and income levels confound any sense of a shared experience, preventing widespread organization (Standing 2011). Instead of solidarity, the precariat feels anxious, alienated, isolated, and angry. It is riddled with internal tensions and its members blame each other, "preventing them from recognizing that the social and economic structure is producing their common set of vulnerabilities" (Standing 2011, p. 25).

Anthropologist Anna Tsing refers to precarity as "[l]ife without the promise of stability" (2015, p. 2). The precariat is characterized by economic uncertainty and debt. Many people in this class work multiple jobs for hourly wages or go from one job to the next with no benefits, security, or chance to accumulate skills and status, often being asked to be on-call for no pay with no guarantee of being called in. They often do work that misaligns with or is perceived as below their level of education or training, leading to inner disapproval and purposelessness. In being on-call or being asked to quickly retool and shift within their jobs, their stress level is high, contributing to higher rates of suicide (Standing 2011). In having to constantly look for new work, search for better work or more work, and promote themselves via their online and/or social interactions, the precariat is all too aware of its instability. Not being paid enough to afford many things associated with American notions success and/or not having the job security to get the benefit of accumulated savings and lower interest, they are continually in debt and pay out much of their income to landlords, lenders, and seller-offered installments on big-ticket items they need or want. They have little hope of accumulating real wealth, and they worry about future poverty even if right now they drive a late model car and own a house.

As Clinton-era welfare reform effectively decreased state aid for many people, applying more and more stringent criteria to qualify, and as poor families face a strong social stigma that blames the less fortunate for not trying hard enough, many will decide to forego aid or only sporadically access it (Morgan et al. 2011; Standing 2018). Whether or not those struggling to access state aid do so, they remain one missed payment away from losing shelter, transportation, utility services, and even going without adequate food. Some have argued that these trends not only correlate but are causative factors in the alarming increase in mass shootings of the past few decades (Berardi 2015; Fraad and Wolff 2017). In attempting to understand and cope with their misfortune, members of the precariat sometimes fall prey to violent vendettas, racial or ethnic scapegoating, fascist ideologies, or populist demagogues, a result we explain below.

The Urban–Rural Divide

While manufacturing declined and the urban precariat grew, rural areas of the U.S. were impacted by the same dynamics of centralization and consolidation. Following World War II, and especially since the 1970s, the economic and social landscape of the U.S. Midwest changed dramatically. America's "breadbasket" moved away from small and medium-sized diversified farms with several crops and animal species towards large-scale monoculture commodity crop farming and meat production using concentrated animal feedlots and centralized processing facilities (MacDonald et al. 2013, 2018). One of the consequences of these changes has been a general depopulation of the countryside and a subsequent concentration of population into urban centers (Johnson and Lichter 2019; McGranahan and Beale 2002).

The psychological literature on displacement and relocation recognizes that forced moves, especially violent and/or sudden ones, cause psychological harm (Erikson 1976; Fullilove 1996). Given the attachments that people form to places – mental, emotional, and cognitive ties often constituting a cultural identity – it is unsurprising that disrupting a particular sense of place results in psychological distress (Tuan 1974, 1977). As Mindy Thompson Fullilove

(2016) writes, "displacement is the problem the twenty-first century must solve. Africans and aborigines, rural peasants and city dwellers have been shunted from one place to another as progress has demanded, 'Land here!' or 'People there!' In cutting the roots of so many people, we have destroyed language, culture, dietary traditions, and social bonds" (p. 5). The concept of "root shock" is useful for understanding the psychic impacts of economic changes in the Midwest. Root shock is "the traumatic stress reaction to the destruction of all or part of one's emotional ecosystem" (Fullilove 2016, p. 11). At the level of the individual, it decimates a person's "working model of the world," leaving them starved of "social, emotional, and financial resources" (p. 14). Root shock at the community level fractures communal bonds, literally ripping apart the social fabric. We can see these effects of dislocation in historic catastrophes such as plague, war, famine and invasion, but also more recent episodes of forced migration, rural and urban development, and the unceasing processes of cultural and economic change. Some react to these types of events by adapting as well as possible (Keyes and Kane 2004). However, when enough people do not, groups can coalesce around violent ideological mass movements (Johnson 2019).

The reason why we do not see much discussion of the effects of economic displacement in the mainstream psychological literature may very well be that, through the lens of pervasive liberal ideology, relocation caused by economic competition and facilitated by democratic political institutions appears voluntary rather than forced (one could always choose to stay, fight and attempt to innovate, and some people do, as we will see in the Rieckmann's story below). Something done "voluntarily" is usually seen as incapable of causing psychological damage. After all, aren't we all free creatures capable of reason (Minogue 1964)? Additionally, due to the pervasive liberal frame, we tend to see ideological movements, even extreme ones, as matters of individual choice rather than as expressions of individual and collective psychological disturbance arising from the structure of society. As Fisher (2008) put it, "It seems that with post-Fordism, the 'invisible plague' of psychiatric and affective disorders … has spread, silently and stealthily since around 1750," and yet "[t]he current ruling ontology denies any possibility of a social causation of mental illness" (pp. 35, 37).

As mentioned briefly above, the changes that have depopulated the rural areas of the United States, especially in the Midwest, were instigated by a slew of government policies that prioritized promoting soft power abroad through free trade in agricultural commodities and encouraged commodity crops through tax policies, subsidies and extension support (Anderson 2014; Cochrane 1993; Danbom 1995; Hurt 2002). Government policies, paired with changes in technology and farm organization, promoted more intensive farming of fewer acres, the consolidation of land into fewer and fewer hands, and the virtual elimination of small, diversified farms, concluding a process that began at least as early as the 1920s with the emergence of the now-dominant industrial ideal in American agriculture (Anderson 2009; Fitzgerald 2003). Small farmers in the Midwest were faced with two very powerful forces – the government and government-aided big agriculture. Prioritizing international markets made the U.S. agricultural sector increasingly influential but also increasingly vulnerable to market shocks. These changes, which had a drastic impact on how and where U.S. citizens lived, were products not of some sort of inevitable evolution but of a history of government decisions in the U.S. and elsewhere, in line with capitalist imperatives, that made agricultural production increasingly analogous to other types of industrial production on the world stage. Figure 3.1 illustrates how dramatic and relatively rapid the transformations were. While these changes have led to increased agricultural productivity, they are also said to "force small farms out of business, damage the viability of rural communities, reduce the diversity of agricultural production, and create environmental risks through their production practices" (MacDonald et al. 2018, p. 1).

(Illinois, Indiana, Iowa, Kansas, Minnesota, Missouri, Nebraska, North Dakota, South Dakota, Wisconsin). In 1940, there were 1,675,297 farms. That number dropped to 605,572 by 2017. In 1940, the average acres per farm was 259; it rose to 656 by 2017.

Hudson and Laingen observed in 2016 that "During the past eight decades, seven out of every ten U.S. farms disappeared" (p. 11). Those that remained had to greatly expand their operations to stay financially afloat, often – irony aside – taking on immense debt in the process (Dudley 2000). The majority of US farms are still family-owned instead of corporate-owned, but the number of farms is far fewer and the size of those that remain is far larger than just a few decades ago (MacDonald et al. 2013). Reflecting on the changes in the scale of American agriculture in the final decades of the 20th

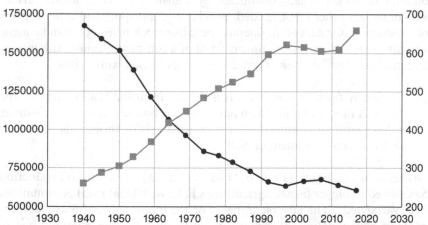

Number of Farms and Average Farm Size (Acres): Midwest, 1940-2017

Figure 3.1 The number of farms and average acres per farm of farms in the Midwest from 1940 to 2017.

Source: Compiled from USDA, NASS Census of Agriculture: 1940–2017.

century, geographer John Fraser Hart (2003) remarked that "farmers have to add a zero or two to the way they once thought, be it dollars or acres, crops or animals, bushels or head" (p. 1). Most farms now focus on 2–3 crops at the most, having invested incredible amounts in specialized machinery, software and proprietary inputs such as seed, fertilizer and pesticide systems that lock farmers into their production choices and make it extraordinarily difficult to independently innovate (Cochrane 1958; MacDonald et al. 2018; Ramey 2010; Stone and Flachs 2017). Livestock operations have moved completely away from growing their own animal food crops and instead purchase feed from other suppliers. According to Stone (2022), the defining feature of contemporary industrial agriculture is that it is completely dependent on external (off-farm) inputs. The result is that the farm operation is something the farmer has to purchase over and over again. Thus, the agricultural sector has taken the same turn as U.S. production generally – toward increasing technical specialization and a concentration of property and profits into ever-fewer hands. However, there is a twist: the concentration of property and profits follows the shift in inputs – it is moving further and further off the farm and out of rural communities (e.g., Davies 1998).

To illustrate just how much farming in the U.S. has been transformed toward consolidation and intensive specialized operations, 4% of farms produce 50% of U.S. agricultural output (Wright 2017). Most farms are still privately owned, but not necessarily privately controlled since farmers in sectors like chicken production are often contractors to corporations, severely limiting farmer autonomy (van der Ploeg 2008). Their owners, especially those who own the largest farms, do not do most of the agricultural work on those farms. Farm owners may not even live on their farms, and if they do, they generally do no more than 20% (with very large farms at no more than 3%) of the farming-related labor (MacDonald et al. 2018). Daily operations on large farms are usually conducted by a farm manager or tenant farmer who organizes other labor needed, purchasing, transport and other aspects of production. Owners often operate more like stock investors than farmers, spending much of their work time in front of a computer dealing with money and market-related matters (Hart 2003). Hart (2003, p. xiii) writes

> the family farm of the 2000s is a far cry from the family farm of the 1980s. A modern family farm has become a business, a very big business indeed, with gross annual returns of hundreds of thousands of dollars and a capital investment of millions.

Drabenstott and Smith (1996, p. 1) write that "[a]s agricultural production has moved to bigger farms, agriculture's links with local rural communities have weakened." The adverse impacts of the industrialization and accompanying consolidation of American agriculture on the communities in which they are located are numerous and well-documented (Lobao 1990; Stofferahn 2006). Industrialized farming is related to lower relative incomes; greater income inequality; higher unemployment and poverty rates; lower total employment generated locally; decreases in local retail trade and diversity of retail firms; declines in population size; increases in crime and civil suits; increases in teenage pregnancies; higher rates of stress and other psychological problems; fading of community organizations and lower rates of participation in community social life; decline in churches and public services; deterioration of water, soil, and air quality; and an abundance of environmental-quality and chemical-related health problems (Blanchette 2020; Buttel and Larson 1979; Goldschmidt 1978; North Central Regional Center for Rural Development 1999; Skees and Swanson 1986; Smithers et al. 2004). Further, communities located near industrial farms experience less democratic decision-making, as agribusiness interests dominate local politics (Goldschmidt 1978; McMillan and Schulman 2003). In short, where industrial farming flourishes, rural lives, livelihoods, and localities disintegrate.

The already massive decline in small farms, most dramatically seen in the Corn Belt and Great Plains, has only gotten more precipitous in the past decade (MacDonald et al. 2018). As Hart (2003) observes, "The old-fashioned, nearly self-sufficient small family farm is a thing of the past. Perhaps the old folks can gradually tighten their belts and still manage to hang on to little one-person farms on land they have inherited, but the younger generation are not willing to make the sacrifices necessary, and they have forsaken the farm in search of a better livelihood and lifestyle" (p. 5).

Hart is referring to stories like that of the Rieckmann family.[6] Featured in *Time Magazine* in 2019, the Rieckmanns' predicament is emblematic of what has transpired for small and medium-sized family farmers as the U.S. agricultural system has transformed (Semuels 2019). Hard-pressed to compete with large-scale mechanized dairy operations with millions of dollars in sales each year and massive government help and subsidization, Mary and John Rieckmann, now both over 80, went into debt to keep the dairy business, which they started in the 1960s, afloat. The Rieckmanns received $16 for every 100 pounds of milk sold, which was 40% less than what they had made even six years prior. As *Time* told its largely urban readership,

> Chapter 12 farm bankruptcies were up 12 percent in the Midwest from July of 2018 to June of 2019; they're up 50 percent in the Northwest. Tens of thousands have simply stopped farming, knowing that reorganization through bankruptcy won't save them. The nation lost more than 100,000 farms between 2011 and 2018; 12,000 of those between 2017 and 2018 alone.
>
> (Semuels 2019)

Unbelievably, the Rieckmanns' farm was still in business in 2021 with the help of a GoFundMe account and local support. A Green Bay NBC news station reported on their still-desperate situation and noted that approximately 500 other Wisconsin farms had gone out of business in 2020 (Bokun 2021). But though they are still in operation, the Rieckmanns are still not making a profit – they just do not know what else to do, and they cannot stop trying because their debt looms large. Their situation characterizes a new normal part of the neoliberal economy. As Deleuze (1992) writes in his "Postscript on the Societies of Control," "Man is no longer man enclosed, but man in debt," always-already striving towards a lower debt-to-asset ratio (p. 6).[7]

Psychological Effects

So far, we have dealt with the near-term trends and issues that have led to our current divisive situation. But the roots of the problem go back farther than the 1970s and Nixon-era agricultural policies. The catastrophic events of the first half of the 20th century were arguably our first foray into mass ideological/political activation in reaction to large-scale economic change, dislocation and social upheaval. In this section, we want to examine that history for clues to what is happening now, particularly the history of enchantment, disenchantment and re-enchantment that characterized the 20th century and still has its grip on us today.

The loss of religious "enchantment" due to the rise of rationalism has been theorized over more than 150 years by many thinkers (e.g., Blumenberg 1985; Critchley 2004; Germain 1993; Horkheimer and Adorno 2004; McCarraher 2020; Nietzsche, 1998, 2000, 2003; Taylor 2007; Voegelin 1952; Weber 1946; but see Bennett 2001). Each in their own way deal with a very important shift in the consciousness of human beings in modernity, a shift with important implications for how we organize ourselves and interact with our environment. Here, we will concentrate on the contributions of Charles Taylor and Carl Jung, authors who focus on the spiritual and psychic aspects of radicalism, and we will combine their perspectives with a further focus on underlying economic causation. These two thinkers are useful because, while they focus on the former (spiritual and psychic), they also incorporate the social, economic, and material dimensions into their analyses as causal factors. All three of these focal points have been downplayed in (neo)liberal ideology, whose skewed emphasis remains on individual agency and moral responsibility.[8]

As more commentators are beginning to point out, (neo)liberal ideology and globalized corporate capitalism have colonized our thinking about matters like religion and psychology that used to be considered separate spheres, not caught up in market logic (Brown 2015). Eugene McCarraher (2020) even argues that as traditional religious faith has waned, we have transferred our enchantments to the all-powerful "Mammon," or the capitalist economy. Global markets obscure pervasive economic exploitation, partly because they encourage a market-oriented way of thinking in almost every area of life (Brown 2015; Foucault 2004). It is now becoming evident that these effects of the capitalist system – psychic/spiritual alienation and pervasive economic exploitation – are interactive and related (see, e.g., Žižek 2012). This may not seem like a novel observation, considering that in the 19th century Marx and Engels (2002) pointed out that "religious fervor" was drowned by capitalism's "icy water of egotistical calculation," and that due to commodification "all that is holy is profaned" (pp. 222–223). Marx and Engels (2002), though, theorized that the drowning of fervor and the profanation of the holy would lead to mass alienation spurring collective proletarian outrage and eventual revolution. Instead, history shows that capitalism and alienation can coexist and even thrive together, and that religious fervor can transfer to ideological radicalization.[9] If this is the case, then the crisis of extreme right-wing populism and terrorism is marked by interdependent ideal and material dimensions, and these must be addressed together.

The far-flung relations of production that span the globe not only obscure the causes of poverty, migrant flows, and environmental threats (Appadurai 1996; Latour 2018; Tsing 2005). Additionally, the social upheavals and environmental wreckage often left in capitalism's wake contribute to an unmooring of self-identity, spiritual impoverishment, and psychological disturbance (Bauman 1996; Berry 2015; Giddens 1991). Such instances of root shock will increase in number and magnitude as the effects of climate change continue to manifest (Bendixson et al. 2020; Berzonsky and Moser 2017). Without investment in spiritual resources, people lose yet another source of worldly stability and security. In the US, this trend plays out in decreasing religious identification among younger generations, including lower rates of attendance and less meaningful participation in religious institutions (Peterson 2018). A lack of religious involvement leaves people with fewer meaningful relationships, sources for collective effort and community resources. As a result, they are more vulnerable to social isolation, unhealthy tribalistic involvements, and irrational mass actions.

The lack of stability in family and community, culture, and place caused by the constant requirement to retool and sell one's labor, incessant changes in communication and production technologies, and ever-present material insecurity in neoliberal capitalism are now becoming too obvious to ignore. Traditional sources of meaning and communal identity like local institutions, churches and other places of worship have been battered not only by their own embroilment in scandal, corruption and general worldliness, but – and arguable more significantly – by the increasing transience of the societies they inhabit. As the sociologist Zygmunt Bauman (2000, 2007) argues, in this era of what he calls "liquid modernity" the former "solids" of social norms and institutions fail to hold themselves together and are melting faster than they can be cast anew. Maintenance of the sources of meaning from which identity is derived requires some form of (cultural) continuity that can withstand the tides of history (Giddens 1991). The precarious present of neoliberal capitalism disallows for any such continuity, leaving its subjects – the precariat – adrift amid seas of uncertainty. From a longer perspective, we shouldn't be surprised that people operating in a globalized economy are losing faith in older social institutions dependent upon culture and place and are exhibiting anxiety, anger, and increasing confusion about where to turn for identity and meaning.[10] In a transient and anxious world, political ideologies and identitarian movements such as white nationalism either replace "church" or embed within "church," becoming the primary source of identity, and nullifying the tangible benefits local institutions could provide (Du Mez 2020).

The Trump supporter who equates Christianity with White nationalism and is sufficiently angry either to storm the Capitol or to support those who did is also part of an economic system whose constant churning has contributed to her feelings of being left behind, even if she is not poor (Ulrich-Schad and Duncan 2018). It also contributes to her sense that she really cannot grasp what is going on politically or economically, which spurs a quest for answers from alternative sources. Global corporate capitalism, conversely, naturally and without necessary intent, thrives on confused and alienated people. It does not, contrary to liberal international relations theory (e.g., Keohane and Nye 1977), need political harmony to thrive, but instead coexists quite well with a great deal of domestic and international division, as long as people keep buying and working (Kirk and Okazawa-Rey 2000). Alienated people who are confused by the dissolution of all they once knew or who are focused only on independently (and competitively) laboring for their next paycheck cannot pose a unified threat to the capitalist system.

Both Carl Jung and Charles Taylor combine observations on declining psychic/spiritual conditions with analyses of long-term environmental, technological and economic changes. Taylor's *A Secular Age* (2007) explains events and trends that led to the repression of religious experience in favor of an "immanent frame" of reference. According to Taylor (2007), the "porous self" characterized in the West by medieval Christianity, the self that had access to direct spiritual experience, was greatly diminished by Enlightenment rationalism. This led to "disenchantment," producing the "buffered self," experienced as a wall between the individual and God and the individual and the collective. For the porous self, "the line between personal agency and impersonal force was not at all clearly drawn," making one vulnerable to the spirits and other more-than-human forces at work in the world (p. 32). Conversely, for the buffered self, that line was much firmer, closing it off to the outside world. Disenchantment and the emergence of the buffered self led to a great "dis-embedding." At this stage, people lost the feeling of oneness with others in their group and experienced themselves as atomistic and autonomous individuals, effectively buffered from the external world. Taylor (2007) argues that dis-embedding necessarily resulted in permanent doubt regarding religion, and this doubt led to an endless quest for meaningful identity. As the loss of social embeddedness increased, the "police state" emerged to make people better Christians who could bridge the divide between God and man through increasing moral purity (Taylor 2007, p. 86). But this was a tragic and dangerous development, because eventually the same methods that were used to produce good Calvinists were unmoored from religious goals and used to create social, political and economic order using the force of the state. People had discovered the power of social engineering.

A century earlier, psychologist Carl Jung (1970a) blamed the catastrophes of World War I and World War II on mass psychosis, identifying the dislocating and disintegrating effects of urbanization, wage-labor, and massification as underlying causes of ideological extremism. Jung (1970b, p. 222) wrote:

> Thanks to industrialization, large portions of the population were up-rooted and were herded together in large centres. This new form of existence – with its mass psychology and social dependence on the fluctuations of markets and wages – produced an individual who was unstable, insecure, and suggestible. He was aware that his life depended on boards of directors and captains of industry, and he supposed, rightly or wrongly, that they were chiefly motivated by financial interests. He knew that, no matter how conscientiously he worked, he could still fall a victim at any moment to economic changes which were utterly beyond his control. And there was nothing else for him to rely on.

Jung (1958) also identified modern empiricism and rationalism as major causes of alienation from the natural world as well as from each other. This alienation made it difficult for people to experience religious faith in psychologically healthy ways. As traditional cultural and religious institutions declined, psychic forces previously channeled through religion and rituals produced an array of neuroses in individuals and psychoses at the societal level. Jung (1966) termed this phenomenon "psychic inflation," which caused people to imagine themselves godlike or to make gods out of charismatic leaders and religions out of mass political movements. In Jung's view, then, the development of destructive mass ideologies like Nazism or totalitarian communism involved something much larger than individuals lacking good reason or proper moral character. Moral accusations like this are still the most likely to be leveled against those who get caught up in contemporary currents of extremism, but Jung thought that all such accusations missed the mark (Jung 2017). In our time, the same psychological dynamics Jung saw in 20th-century fanaticism can help us explain the uncanny mass ideological attachments to the person of Donald Trump that led to the Capitol insurrection, and continue in the form of Trump rallies despite his losing of the 2020 election and their risk of being Covid-19 super-spreader events.

Resonating with concerns that are still very relevant today, Jung argued that uniform, routinized work and increasing urbanization created an unhealthy "herd" mentality. As massification proceeded, moral responsibility became more diffused, promoting the excuse of "just following orders" (Balfour et al. 2020). The most well-known and tragic example of this dynamic is given by the case of Adolf Eichmann, who played a major role in the Nazi's "Final Solution," and personally arranged for the transport of hundreds of thousands of Jews to extermination camps. While on trial for his crimes in Jerusalem, Eichmann pleaded with the court again and again that he was simply following orders, doing his duty (Arendt 1994). Thus, the Eichmann case reveals the dangers of being unable to think for one's self: "one doesn't need to be fanatical, sadistic, or mentally ill to murder millions; ... it is enough to be a loyal follower eager to do one's duty" (Simon Wiesenthal quoted in Levy (1993, pp. 157–158)).

Jung (1968) argued that people who did not experience massification were more aware of their individuality and moral agency (p. 131). If they worked on farms, were craftsmen, or engaged in local business, they could more easily retain a sense of agency and responsibility, which they could express in a concrete way due to their proximity to extended family. Because of this, so long as they had enough people, their communities were stronger than those in urban areas. As the opportunity to continue this rural way of life dwindled, a culture clash was in the making between the growing number of people who lived in urban areas and the decreasing number of non-urban citizens whose way of life was being eroded and whose way of thinking was considered by the larger urbanized society to be retrograde.

Jung (1966) observed that the rewards of urban areas (wages) were precarious in ways that the rewards of subsistence living were not. In a pre-industrial setting, an agricultural worker's livelihood might have been affected by "acts of God" such as fire, drought or flood; in a city, a worker's prospects were profoundly affected by the faceless "they." Worse yet, the (purportedly autonomous) market was an entirely impersonal Behemoth they could not get around (Jung 1970b). To the extent that governments intervened in the economy, and they most certainly did in all the economic systems known at the time, they did so not in the interest of individuals, families or groups but rather with abstract technological progress and economic growth in mind.

The observations of Jung and Taylor outlined above resonate with the earlier concerns of Karl Marx, who, with Friedrich Engels (2002), observed in the 19th century that capitalism "has resolved personal worth into exchange value, and in place of the numberless indefeasible chartered freedoms, has set up that single, unconscionable freedom – Free Trade" (p. 222). Jung (1958, 1970a, 1970b) noted the tendency of the state, in response to the damage done by the impersonal market, to intervene in the capitalist economy in regulating, planning and organizing many areas of life – banking, economic development, farming and food, housing, education and healthcare. All of this intervention was done in the name of an economic system that promised a certain type of consumer "freedom" (Harvey 2005; Konings 2018). As people in liberal capitalist countries became more dependent upon markets *and* governments for their survival, they also became vulnerable to a kind of mass ideological mentality, albeit mostly less rabid that that exhibited recently at the "Unite the Right" Charlottesville riot and the insurrection at the Capitol.

In addition to a sense of precarity and dependency that produced profound anxiety, part of the price of industrialized, technological civilization was a deadening of the spiritual senses and a diminishment of the feeling of being alive. Following Italian philosopher Franco "Bifo" Berardi (2005), we can refer to this dynamic as the "dis-eroticization of daily life," wherein "one finds less and less pleasure and less and less reassurance" (p. 60). The social, cultural and economic forces that caused this dehumanization, disenchantment, or dis-eroticization – and that subsequently caused people to throw themselves into extreme ideologies as a means of compensation – are still with us today and have emerged in right-wing populism, ethno-nationalism, and various other highly ideological identifications that can brook no compromises. At the deepest level, the forces that cause these disturbances include scientism, automation, capitalism, globalization, and technical rationality.

Since the 1940s, the rural/urban divide has simply gotten worse. Alcohol and drug addiction are a huge problem in small-town America (Reding 2009). Rural citizens lack access to good medical care and often reside in food deserts (Blanchard and Matthews 2007; Douthit et al. 2015; Morton et al. 2005). Ironically, rural and inner-city areas share similar problems, and both are often forgotten, treated by the rest of society as the detritus left behind by economic progress (Fullilove 2016). Wendell Berry (2015) foresaw the consequences of this great upheaval when he compared the demolishment of white rural communities to the destruction of Native Americans. The two are not to be equated, but the same dynamic was behind both destructions: the "conquerors" destroying settled ways of life in the name of inevitable, unstoppable economic "progress." Berry (2015, p. 6) lamented:

> Time after time, in place after place, these conquerors have fragmented and demolished traditional communities, the beginnings of domestic cultures. They have always said that what they destroyed was outdated, provincial, and contemptible. And with alarming frequency they have been believed and trusted by their victims, especially when their victims were other white people.

As the U.S. moved away from the agrarian life, traditional means of independence were destroyed, and spiritual resources such as churches and civic/ service organizations disappeared from the landscape. Contrary to what many in the United States are told, this decline is not the fault of "big government," the "nanny state," or "cultural Marxism." Instead, the ultimate origins of this decline are deeper and therefore longer in the making – in the scientific revolution, the Enlightenment, and the urbanizing, industrializing, and globalizing processes of capitalist development.

Conclusions

While voting, especially in presidential election years, is still an important aspect of U.S. political participation (and was on the upswing in the 2020 faceoff between Trump and Biden), the political bandwidth of a growing number of Americans has been taken up by largely mediated identity movements (Erbschloe 2021). This is a subset of a larger phenomenon – the majority of Americans are online every day and increasingly experience their social and economic life through social media, TV, internet news and online services (Shearer and Mitchell 2021). Trying to get a handle on what happened post-election at the U.S. Capitol, many have voiced the opinion that the pandemic created the conditions that led to over-reliance on Internet sources of news and socialization, and that created alternative realities such as Q-Anon, militia movements, etc. But, of course, these types of movements preceded the pandemic and have a long trajectory – they just recently moved thoroughly into the mainstream media purview and the consciousness of even those who pay little attention to the news. They are also part of a global phenomenon, not something that is just isolated to the U.S., as evidenced by the election of Bolsonaro in Brazil, Erdoğan in Turkey, and the growing popularity of nationalist and other far-right parties in European and other countries (BBC 2019). We have argued above that they are the result of larger underlying economic changes that have left the majority of people in many countries in a precarious situation. However, there is no doubt that our increasingly mediated existence has exacerbated radicalization, both mild and strong, in response to the fear and anxiety caused by rapid economic changes and growing precarity (Erbschloe 2021; Krzych 2021; Nagle 2017).

The past few decades have been marked by an insidious effect of our media-infused lives – the confusion of political opinionating with political action. Whether we call it "bread and circuses" (or, in the case of Trump, "bread and tweets" [Roubini 2017]) or, as Guy Debord (1995) put it, "the Spectacle," almost everyone in the U.S. either fully or at least partly lives in a world in which reality is on the screen, and what motivates people has more to do with identity and felt social acceptability than material reality (Briziarelli and Armano 2017; Hedges 1994). In this environment, and despite even conservative warnings, it is no wonder that an obvious narcissist could be elected president and be loved by millions despite his seeming disregard for their dignity and wellbeing (Swogger and Miller 2016). As Kellner (2017, p. 1) recently argued,

> We now live in an era, where the digitally mediated spectacle has contributed to right-wing authoritarian populist Donald Trump becoming U.S. president, and Debord's concept of spectacle is now more relevant than ever to interpreting contemporary culture, society, and politics.

The problem with living in the Spectacle is not just that individuals lack authenticity and develop shallow tastes in entertainment. Social media is designed to make people feel as if they are participating in something real and being heard, even though they are not. When it comes to politics, the effect of this great diversion into the digital is crippling. As more and more feel as though they have taken political action with a tweet, Facebook post, YouTube video, video comment, blog post, or simply by insisting on watching only one news network, they fall into a trap that for the most part channels their political energies into feelings-based, non-productive and largely narcissistic virtue-signaling rather than real concerted action, as we saw in the mass events of 2020 such as occupations like Capitol Hill Autonomous Zone (now Capitol Hill Occupied Protest) in Seattle Black Lives Matter street protests, riots, marches and, of course, the Insurrection on the Capitol, when online-augmented discontents do erupt into real political action, the action is uncoordinated, prone to violence, and short-lived, followed by no real positive change or even further regression.[11]

Our associational and institutional means of direct participation are becoming incoherent and sclerotic largely because of the dynamics we have discussed above: our changed economy that makes people insecure and anxious, removes a sense of belonging and identity, fragments society and pits people against each other. But the efficient cause of our times is the fact that it is much more immediately satisfying to tweet. Slavoj Žižek (2012) rightly points out that people in Western societies want to enjoy without paying the price – in this case, to enjoy the catharsis without effecting any change. He also argues that explosions of political mass action and even violence can be positive, but only if such things are not simply pressure releases but lead to effective action in the direction of fixing the problems causing the explosion (Žižek 2009b).

The fact that people most often cannot effectively act in a collective manner and either are content to burn up endless hours online or sometimes spill out into performative but real-world demonstrations is the result, according to Žižek (2012), of the particular effects of global capitalism. We literally live in a world without collective meaning, in an order that has created the perfect conditions to prevent agreement and action on any such meaning. Zizek (2012, p. 55) writes:

> Perhaps it is here that we should locate one of the main dangers of capitalism. Although capitalism is global, encompassing the whole world, it sustains a stricto sensu "worldless" ideological constellation, depriving the vast majority of people of any meaningful cognitive orientation. Capitalism is the first socio-economic order which de-totalizes meaning: it is not global at the level of meaning. There is, after all, no global "capitalist worldview," no "capitalist civilization" proper. The fundamental lesson of globalization is precisely that capitalism can accommodate itself to all civilizations… Capitalism's global dimension can only be formulated at the level of truth-without-meaning, as the real of the global market mechanism.

In this situation, people will grasp for meaning as individuals and look for tribes to provide it, but without effect. They mistake their flights into ideology with belonging, and most often achieve nothing more positive than consuming more products. After all, the purpose of Fox News is advertising.

Notes

1 The sheer cartoonish and cringeworthy nature of many of these media outlets and personalities should have been the first clue that we were dealing with some sort of cultism, but until recently, most mainstream analysts remained unacquainted with the nature of their communication and tropes. Angela Nagle's *Kill All Normies* (Nagle 2017) was groundbreaking in this sense, as it explores the "dark corners of the anonymous internet" where the "openly white nationalist alt-right"

was able to organize and come into its own, all while remaining hidden among an "online army of ironic in-jokey trolls." As she writes: "Before the overtly racist alt-right were widely known, the more mainstream alt-light largely flattered it, gave it glowing write-ups in Breitbart and elsewhere, had its spokespeople on their YouTube shows and promoted them on social media" (pp. 8–9). For older social-psychological analyses of cultish right-wing movements, see Adorno et al. (2019), Freud (1975), Reich (1970).

2 In Chapter 15 of the first volume of *Capital*, titled "Machinery and Large Scale Industry," Marx (1976) writes: "Like every other instrument for increasing the productivity of labour, machinery is intended to cheapen commodities and, by shortening the part of the working day in which the worker works for himself, to lengthen the other part, the part he gives to the capitalist for nothing. The machine is a means for producing surplus-value" (p. 492). See also Marx's "Fragment on Machines" in the *Grundrisse* (Marx 1973, pp. 690–712), where he notes that automation produces alienation among laborers.

3 Wark (2020) finds this effect so pronounced that she even wonders if we've entered a post-capitalist phase in which owners of information vectors rule over the relatively-stuck owners of capital.

4 The privatization of the commons is referred to across anti-capitalist literature as enclosure. See Marx (1976, Part VIII) for the original formulation of the enclosure and Hardt and Negri (2000) for a contemporary deployment. For a contemporary treatment of the concept of commons, see Ostrom (1990).

5 The United Nations' World Social Report 2020: *Inequality in a Rapidly Changing World*, states that while inequality across countries has actually declined in the past few decades, inequality within countries, particularly developed countries, is on the rise (Department of Economic and Social Affairs 2020).

6 For additional stories in a similar vein, see Dudley (2000).

7 Maurizo Lazzarato (2012) argues that "Debt constitutes the most deterritorialized and the most general power relation through which the neoliberal power bloc institutes its class struggle" (p. 89). Hannah Appel et al. (2019) see mass indebtedness as potential for collective action.

8 Wendy Brown (2015) terms this emphasis on responsibility "responsibilization" (p. 84)

9 On the coexistence of capitalism and alienation, see Horkheimer and Adorno (2004), McGowan (2016), and Žižek (2009a). On ideological radicalization, see Johnson (2019).

10 Fisher (2008) points out the way in which the system creates depression and anxiety and then treats those who have these conditions as though the problem is simply individual chemical imbalances, thus obscuring the systemic causes and perpetuating them.

11 We understand political regression as no positive movement in the balance of power between labor and corporate power. In other words, political regression is no positive change in the autonomy of the proletarianized masses.

References

Aaronson, D. and Mazumder, B. 2008. Intergenerational economic mobility in the United States, 1940 to 2000. *Journal of Human Resources* 43(1):139–172.

Adorno, T., Frenkel-Brunswick, E., Levinson, D., and Sanford, R. 2019. *The Authoritarian Personality*. New York: Verso Books.

Aiginger, K. 2020. Populism: Root causes, power grabbing and counter strategy. *Intereconomics* 55(1):38–42.

Anderson, J. 1991. The Carter Administration and regulatory reform: Searching for the right way. *Congress and the Presidency* 18(2):121–146.

Anderson, J. 2009. *Industrializing the Corn Belt: Agriculture, Technology and Environment, 1945–1972*. DeKalb, IL: Northern Illinois University Press.

Anderson, J. 2014. Uneasy dependency: Rural and farm policy in the Midwest since 1945. In J. Anderson (ed.) *The Rural Midwest Since WW2*. pp. 126–159. DeKalb, IL: Northern Illinois University Press.

Appadurai, A. 1996. *Modernity at Large: Cultural Dimensions of Globalization*. Minneapolis: University of Minnesota Press.

Appel, H., Whitley, S., and Kline, C. 2019. *The Power of Debt: Identity and Collective Action in the Age of Finance*. https://escholarship.org/uc/item/2hc1r7fx.

Arendt, H. 1994. *Eichmann in Jerusalem: A Report on the Banality of Evil*. New York: Penguin Books.

Aronowitz, S. 2014. *The Death and Life of American Labor: Toward a New Workers' Movement*. New York: Verso Books.

Balfour, D., Adams, G., and Nickels, A. 2020. *Unmasking Administrative Evil* (Fifth ed.). New York: Routledge.

Bauman, Z. 1996. From pilgrim to tourist – Or a short history of identity. In S. Hall and P. Du Gay (eds.) *Questions of Cultural Identity*. pp. 18–36. London: SAGE.

Bauman, Z. 2000. *Liquid Modernity*. Cambridge: Polity Press.

Bauman, Z. 2007. *Liquid Times: Living in an Age of Uncertainty*. Cambridge: Polity Press.

Beckert, J. 2016. *Imagined Futures: Fictional Expectations and Capitalism Dynamics*. Cambridge, MA: Harvard University Press.

Bendixson, C., Durbin, T., and Hanschu, J. 2020. "'Progressive ranching" and wrangling the wind as ecocultural identity maintenance in the Anthropocene. In T. Milstein and J. Castro-Sotomayor (eds.) *Routledge Handbook of Ecocultural Identity*. pp. 164–178. New York: Routledge.

Bennett, J. 2001. *The Enchantment of Modern Life: Attachments, Crossings, and Ethics*. Princeton, NJ: Princeton University Press.

Berardi, F. 2015. *Heroes: Mass Murder and Suicide*. New York: Verso.

Berry, W. 2015. *The Unsettling of America: Culture and Agriculture*. Berkeley: Counterpoint Press.

Berzonsky, C. and Moser, S. 2017. Becoming homo sapiens sapiens: Mapping psycho-cultural transformation in the Anthropocene. *Anthropocene 20*:15–23.

Blanchard, T. and Matthews, T. 2007. Retail concentration, food deserts, and food-disadvantaged communities in rural America. In C. Hinrichs and T. Lyson (eds.) *Remaking the North American Food System: Strategies for Sustainability*. pp. 201–215. Lincoln, NE: University of Nebraska Press.

Blanchette, A. 2020. *Porkopolis: American Animality, Standardized Life, and the Factory Farm*. Durham: Duke University Press.

Blumenberg, H. 1985. *The Legitimacy of the Modern Age*. (R. Wallace, trans.) Cambridge: MIT Press.

Bokun, B. 2021. As 500 Wisconsin farmers disappear, a Fremont family battles to keep theirs running amid declining sales. *NBC 26: Green Bay* (Feb. 22). https://www.nbc26.com/news/local-news/the-number-of-small-family-owned-farms-in-wisconsin-continues-to-decline.

Briggs, W. 2021. *A Cauldron of Anxiety: Capitalism in the Twenty-First Century*. Winchester, UK: Zero Books.

British Broadcasting Company (BBC). 2019, November 13). Europe and right-wing nationalism: A country-by-country guide. https://www.bbc.com/news/world-europe-36130006.

Briziarelli, M. and Armano, E. (eds.) 2017. *The Spectacle 2.0: Reading Debord in the Context of Digital Capitalism*. Westminster: University of Westminster Press.

Brown, J. 2018. *Millennials and Retirement: Already Falling Short*. https://www.nirsonline.org/wp-content/uploads/2018/02/Millennials-Report-1.pdf.

Brown, W. 2015. *Undoing the Demos: Neoliberalism's Stealth Revolution*. New York: Zone Books.

Buttel, F. and Larson III, O. 1979. Farm size, structure, and energy intensity: An ecological analysis of U.S. agriculture. *Rural Sociology 44*:471–488.

Chien, Y. and Morris, P. 2018. Accounting for age: The financial health of millennials. *Regional Economist* (Second Quarter). https://www.stlouisfed.org/publications/regional-economist/second-quarter-2018/accounting-age-financial-health-millennials.

Cochrane, W. 1958. *Farm Prices, Myth and Reality*. Minneapolis: University of Minnesota Press.

Cochrane, W. 1993. *The Development of American Agriculture: A Historical Analysis* (Second ed.). Minneapolis: University of Minnesota Press.

Crary, J. 2013. *24/7: Late Capitalism and the Ends of Sleep*. London: Verso Books.

Critchley, S. 2004. *Very Little … Almost Nothing: Death, Philosophy, Literature* (Second ed.). New York: Routledge.

Danbom, D. 1995. *Born in the Country: A History of Rural America*. Baltimore: Johns Hopkins University Press.

Davies, R. 1998. *Main Street Blues: The Decline of Small-Town America*. Columbus: Ohio State University Press.

Debord, G. 1995. *The Society of the Spectacle*. (D. Nicholson-Smith, trans.) New York: Zone Books.

Deleuze, G. 1992. Postscript on the societies of control. *October* 59:3–7.

Department of Economic and Social Affairs. 2020. *World Social Report 2020: Inequality in a Rapidly Changing World*. https://www.un.org/development/desa/dspd/world-social-report/2020-2.html.

Dettling, L. and Hsu, J.W. 2017. Playing catch-up. *Finance and Development 54*(2). https://www.imf.org/external/pubs/ft/fandd/2017/06/dettling.htm.

Dimitri, C., Effland, A., and Conklin, N. 2005. *The 20th Century Transformation of U.S. Agriculture and Farm Policy*. Economic Research Service, Economic Information Bulletin 3. https://www.ers.usda.gov/webdocs/publications/44197/13566_eib3_1_.pdf?v=5016.5.

Douthit, N., Kiv, S., Dwolatzky, T., and Biswas, S. 2015. Exposing some important barriers to health care access in the rural USA. *Public Health 129*:611–620.

Drabenstott, M. and Smith, T. 1996. The changing economy of the rural heartland. In Federal Reserve Bank of Kansas City (ed.) *Economic Forces Shaping the Rural Heartland, pp. 1-11*. Kansas City: Federal Reserve Bank.

Dudley, K.M. 2000. *Debt and Dispossession: Farm Loss in America's Heartland*. Chicago: University of Chicago Press.

Du Mez, K. 2020. *Jesus and John Wayne: How White Evangelicals Corrupted a Faith and Fractured a Nation*. New York: Liveright.

Ehrenreich, B. 2020. *Fear of Falling: The Inner Life of the Middle Class*. New York: Twelve Press.

Erbschloe, M. 2021. *Extremist Propaganda in Social Media: A Threat to Homeland Security*. New York: CRC Press.

Erikson, K. 1976. *Everything in Its Path: Destruction of Community in the Buffalo Creek Flood*. New York: Simon and Schuster.

Fisher, M. 2008. *Capitalist Realism: Is There No Alternative?* Winchester, UK: Zero Books.

Fitzgerald, D. 2003. *Every Farm a Factory: The Industrial Ideal in American Agriculture*. New Haven: Yale University Press.

Foltz, J., Jackson-Smith, D., and Chen, L. 2002. Do purchasing patterns differ between large and small dairy farms? Econometric evidence from three Wisconsin communities. *Agricultural and Resource Economics Review 31*(1):28–38.

Foucault, M. 2004. *The Birth of Biopolitics: Lectures at the College de France, 1978–1979*. (G. Burchell, trans.) New York: Picador.

Fraad, H and Wolff, R. 2017. American hyper-capitalism breeds the lonely, alienated menwhobecomekillers.*Salon*(Nov.8).https://www.salon.com/2017/11/08/american-hyper-capitalism-breeds-the-lonely-alienated-men-who-become-mass-killers_partner/.

Freud, S. 1975. *Group Psychology and the Analysis of the Ego*. (J. Strachey, trans.) New York: W.W. Norton.

Fullilove, M. 1996. Psychiatric implications of displacement: Contributions from the psychology of place. *American Journal of Psychiatry 153*(12):1516–1523.

Fullilove, M. 2016. *Root Shock: How Tearing Up City Neighborhoods Hurts America, and What We Can Do About It*. (Second ed.). New York: New Village Press.

Gardner, B. 2002. *American Agriculture in the Twentieth Century: How It Flourished and What It Cost*. Cambridge, MA: Harvard University Press.

Germain, G. 1993. *A Discourse on Disenchantment: Reflections on Politics and Technology*. Albany, NY: State University of New York Press.

Giddens, A. 1991. *Modernity and Self-Identity: Self and Society in the Late Modern Age*. Redwood City, CA: Stanford University Press.

Goldschmidt, W. 1978. *As You Sow: Three Studies in the Social Consequences of Agribusiness*. Montclair, NJ: Allanheld, Osmun and Company.

Hardt, M. and Negri, A. 2000. *Empire*. Cambridge, MA: Harvard University Press.

Hart, J. 2003. *The Changing Scale of American Agriculture*. Charlottesville, VA: University of Virginia Press.

Harvey, D. 2005. *A Brief History of Neoliberalism*. Oxford: Oxford University Press.

Hedges, C. 1994. *Empire of Illusion: The End of Literacy and the Triumph of Spectacle*. (Eighth ed.). New York: Bold Type Books.

Hendrickson, M., Howard, P. and Constance, D. 2019. Power, food and agriculture: Implications for farmers, consumers and communities. In J. Gibson and S. Alexander (eds.) *In Defense of Farmers: The Future of Agriculture in the Shadow of Corporate Power*. pp. 13–61. Lincoln, NE: University of Nebraska Press.

Horkheimer, M. and Adorno, T. 2004. *Dialectic of Enlightenment: Philosophical Fragments*. (E. Jephcott, trans.) Stanford: Stanford University Press.

Howard, P. 2021. *Concentration and Power in the Food System: Who Controls What We Eat?* (Revised ed.). London: Bloomsbury Academic.

Hudson, J. and Laingen, C. 2016. *American Farms, American Food: A Geography of Agriculture and Food Production*. New York: Lexington Books.

Hurt, R. 2002. *American Agriculture: A Brief History*. (Revised ed.). West Lafayette, IN: Purdue University Press.

Johnson, K. and Lichter, D. 2019. Rural depopulation: Growth and decline processes over the past century. *Rural Sociology 84*(1):3–27.

Johnson, L. 2019. *Ideological Possession and the Rise of the New Right: The Political Thought of Carl Jung*. New York: Routledge.

Jung, C. 1958. *The Undiscovered Self*. New York: Mentor Books.

Jung, C. 1966. On the psychology of the unconscious. In G. Adler (ed.) *The Collected Works of C.G. Jung, Volume 7: Two Essays on Analytical Psychology*. (Second ed). pp. 1–121. (R. Hull, trans.) Princeton: Princeton University Press.

Jung, C. 1968. Concerning rebirth. In G. Adler (ed.) *The Collected Works of C.G. Jung, Volume 9 (Part 1): The Archetypes and the Collective Unconscious*. (Second ed). pp. 113–150. (R. Hull, trans.). Princeton: Princeton University Press.

Jung, C. 1970a. After the catastrophe. In G. Adler (ed.) *The Collected Works of C.G. Jung, Volume 10: Civilization in Transition*. (Second ed.). pp. 194-217. (R. Hull, trans.) Princeton: Princeton University Press.

Jung, C. 1970b. The fight with the shadow. In G. Adler (ed.) *The Collected Works of C.G. Jung, Volume 10: Civilization in Transition*. (Second ed). pp. 218–226. (R. Hull, trans.). Princeton: Princeton University Press.

Jung, C. 2017. *Modern Man in Search of a Soul*. (W. Dell and C. Baynes, trans.). Eastford, CT: Martino Fine Books.

Kellner, D. 2017. Guy Debord, Donald Trump, and the politics of spectacle. In M. Briziarelli and E. Armano (eds.). *The Spectacle 2.0: Reading Debord in the Context of Digital Capitalism*. pp. 1–13. Westminster: University of Westminster Press.

Kelo v City of New London et al. 2005. 545 U.S. no. 04–108 (Supreme Court of the United States).

Keohane, R. and Nye, J. 1977. *Power and Interdependence: World Politics in Transition*. Boston: Little, Brown and Company.

Keyes, E. and Kane, C. 2004. Belonging and adapting: Mental health and Bosnian refugees living in the United States. *Issues in Mental Health Nursing* 25(8):809–831.

Kirk, G. and Okazawa-Rey, M. 2000. Neoliberalism, militarism, and armed conflict. *Social Justice* 27(4):1–17.

Konings, M. 2018. *Capital and Time: For a New Critique of Neoliberal Reason*. Stanford: Stanford University Press.

Krzych, S. 2021. *Beyond Bias: Conservative Media, Documentary Form, and the Politics of Hysteria*. Oxford: Oxford University Press.

Lasch, C. 1996. *The Revolt of the Elites and the Betrayal of Democracy*. New York: W.W. Norton.

Latour, B. 2018. *Down to Earth: Politics in the New Climatic Regime*. Cambridge: Polity Press.

Lazzarato, M. 2012. *The Making of Indebted Man: An Essay on the Neoliberal Condition*. (J. Jordan, trans). Los Angeles: Semiotext(e).

Levy, A. 1993. *Nazi Hunter: The Wiesenthal File*. London: Constable and Robinson.

Lobao, L. 1990. *Locality and Inequality: Farm and Industry Structure and Socioeconomic Condition*. Albany, NY: State University of New York Press.

Lyson, T., Torres, R., and Welsh, R. 2001. Scale of agricultural production, civic engagement, and community welfare. *Social Forces* 80:311–327.

MacDonald, J., Hoppe, R., and Newton, D. 2018. *Three Decades of Consolidation in U.S. Agriculture*. Economic Research Service, Economic Information Bulletin 189. https://www.ers.usda.gov/webdocs/publications/88057/eib-189.pdf?v=792.6.

MacDonald, J., Korb, P., and Hoppe, R. 2013. *Farm Size and the Organization of U.S. Crop Farming*. Economic Research Service, Economic Research Report 152. https://www.ers.usda.gov/webdocs/publications/45108/39359_err152.pdf

Marx, K. 1973. *Grundrisse: Foundations of the Critique of Political Economy*. (M. Nicolaus, trans). London: Penguin Books.

Marx, K. 1976. *Capital: A Critique of Political Economy, Volume 1*. (B. Fowkes, trans.) London: Penguin Books.

Marx, K. and Engels, F. 2002. *The Communist Manifesto*. New York: Penguin Classics.

McCarraher, E. 2020. *Enchantments of Mammon: How Capitalism Became the Religion of Modernity*. Cambridge: Belknap Press.

McGowan, T. 2016. *Capitalism and Desire: The Psychic Cost of Free Markets*. New York: Columbia University Press.

McGranahan, D. and Beale, C. 2002. Understanding rural population loss. *Rural America 17*(4):2–11.

McMillan, M. and Schulman, M. 2003. Hogs and citizens: A report from the North Carolina front. In W. Falk, M. Schulman, and A. Tickamyer (eds.) *Rural Restructuring in Local and Global Contexts*. pp. 219–239. Athens, OH: Ohio University Press.

Minogue, K. 1964. *The Liberal Mind: A Critical Analysis of the Philosophy of Liberalism and its Political Effects*. New York: Vintage Books.

Morgan, S., Acker, J., and Weigt, J. 2011. *Stretched Thin: Poor Families, Welfare Work, and Welfare Reform*. Ithaca, NY: Cornell University Press.

Morton, L., Bitto, E., Oakland, M., and Sand, M. 2005. Solving the problems of Iowa food deserts: Food insecurity and civic structure. *Rural Sociology 70*(1):94–112.

Nagle, A. 2017. *Kill All Normies: Online Culture Wars from 4chan and Tumblr to Trump and the Alt-Right*. London: Zero Books.

Nakamura, D. 2015. Obama defends free trade push to supporters: This isn't NAFTA. *Washington Post* (April 23). https://www.washingtonpost.com/news/post-politics/wp/2015/04/23/obama-defends-free-trade-push-to-supporters-this-isnt-nafta/.

Nietzsche, F. 1998. *Beyond Good and Evil: A Prelude to a Philosophy of the Future*. (M. Faber, trans.) Oxford: Oxford University Press.

Nietzsche, F. 2000. *The Birth of Tragedy*. (D. Smith, trans.) Oxford: Oxford University Press.

Nietzsche, F. 2003. *Thus Spake Zarathustra: A Book for All and None*. New York: Algora.

North Central Regional Center for Rural Development (NCRCRD). 1999. *The Impact of Recruiting Vertically Integrated Hog Production in Agriculturally-Based Counties of Oklahoma*. Report to the Kerr Center for Sustainable Agriculture. Ames, IA: Iowa State University.

Ostrom, E. 1990. *Governing the Commons: The Evolution of Institutions for Collective Action*. Cambridge: Cambridge University Press.

Paarlberg, R. and Paarlberg, D. 2000. Agricultural policy in the twentieth century. *Agricultural History 74*(2):136–161.

Peterson, P. (ed.) 2018. *The Decline of Established Christianity in the Western World: Interpretations and Responses*. New York: Routledge.

Piketty, T. 2014. *Capital in the Twenty-First Century*. (A. Goldhammer, trans.) Cambridge, MA: Belknap Press.

Ramey, E. 2010. Seeds of Change: Hybrid Corn, Monopoly, and the Hunt for Superprofits. *Review of Radical Political Economics* 42(3):381–386.

Reding, N. 2009. *Methland: The Death and Life of an American Small Town*. New York: Bloomsbury.

Reich, W. 1970. *The Mass Psychology of Fascism*. New York: Farrar, Straus and Giroux.

Roubini, N. 2017. Trump governs like Nero, appeasing the masses with bread and tweets. *MarketWatch* (Dec. 11). https://www.marketwatch.com/story/trump-governs-like-nero-appeasing-the-masses-with-bread-and-tweets-2017-12-11.

Sautter, J., Furrey, L., and Gresham, R. 2007. Construction of fool's paradise: Ethanol subsidies in America. *Sustainable Development Law and Policy* 7(3):26–29.

Semuels, A. 2019. "They're trying to wipe us off the map." Small American farmers are nearing extinction. *Time* (Nov. 27). https://time.com/5736789/small-american-farmers-debt-crisis-extinction/.

Shearer, E. and Mitchell, A. 2021. News use across social media platforms in 2020. *Pew Research Center Report*. https://apo.org.au/node/311092.

Skees, J. and Swanson, L. 1986. Examining policy and emerging technologies affecting farm structure in the South and the interaction between farm structure and well-being of rural areas. In Office of Technology Assessment, U.S. Congress (ed.) *Public Policy and the Changing Structure of American Agriculture*. pp. 373–495. Washington, DC: Office of Technology Assessment.

Smithers, J., Johnson, P., and Joseph, A. 2004. The dynamics of family farming in North Huron County, Ontario. Part II: Farm–community interactions. *Canadian Geographer* 48:209–224.

Standing, G. 2011. *The Precariat: The New Dangerous Class*. London: Bloomsbury.

Standing, G. 2018. *The Corruption of Capitalism: Why Rentier Thrive and Work Does Not Pay*. London: Biteback Publishing.

Stofferahn, C. 2006. *Industrialized Farming and Its Relationship to Community Well-Being: An Update of a 2000 Report by Linda Lobao*. Report prepared for the State of North Dakota, Office of the Attorney General.

Stone, G. 2022. *The Agricultural Dilemma: How Not to Feed the World*. New York: Routledge.

Stone, G. and Flachs, A. 2017. The ox fall down: Path-breaking and technology tread-mill in Indian cotton agriculture. *Journal of Peasant Studies* 45:1272–1296.

Swogger Jr., G. and Miller, H. 2016. Donald Trump: Narcissist-in-chief, not Commander-in-chief. *Forbes Capital Flows* (March 30). https://www.forbes.com/sites/realspin/2016/03/30/donald-trump-narcissist-in-chief-not-commander-in-chief/?sh=2ab6aa185d45.

Taylor, C. 2007. *A Secular Age*. Cambridge, MA: Belknap Press.

Tsing, A. 2005. *Friction: An Ethnography of Global Connection*. Princeton: Princeton University Press.

Tsing, A. 2015. *Mushroom at the End of the World: On the Possibility of Life in Capitalist Ruins*. Princeton: Princeton University Press.

Tuan, Y. 1974. *Topophilia: A Study of Environmental Perception*. Columbia: Columbia University Press.

Tuan, Y. 1977. *Space and Place: The Perspective of Experience*. Minneapolis: University of Minnesota Press.

Ulrich-Schad, J. and Duncan, C. 2018. People and places left behind: Work, culture, and politics in the rural United States. *Journal of Peasant Studies* 45(1):59–79.

Van der Ploeg, J. 2008. *The New Peasantries: Struggles for Autonomy and Sustainability in an Era of Empire and Globalization*. London: Earthscan.

Voegelin, E. 1952. *The New Science of Politics: An Introduction*. Chicago: University of Chicago Press.

Wallander, S., Claassen, R., and Nickerson, C. 2011. *The Ethanol Decade: An Expansion of U.S. Corn Production, 2000–2009*. Economic Research Service, Economic Information Bulletin 79. https://www.ers.usda.gov/publications/pub-details/?pubid=44566.

Wark, M. 2020. *Capital Is Dead: Is This Something Worse?* London: Verso Books.

Weber, M. 1946. Science as vocation. In H. Gerth and C.W. Mills (eds. and trans.) *From Max Weber: Essays in Sociology.* pp. 129–156. New York: Oxford University Press.

Williams, J. 2006. The pedagogy of debt. *College Literature 33*(4):155–169.

Wright, K. 2017. Agricultural consolidation causes and the path forward: The 2017 Agricultural Symposium. *Ten Magazine.* Federal Reserve Bank of Kansas City. https://www.kansascityfed.org/ten/2017-fall-ten-magazine/agsymposium/.

Zakaria, F. 2016. What's really pushing politics to the right? Immigration: The West needs to do a better job of helping foreigners assimilate. *Washington Post* (Dec. 8). https://www.washingtonpost.com/opinions/whats-really-pushing-politics-to-the-right-immigration/2016/12/08/7a7553c8-bd8a-11e6-91ee-1adddfe36cbe_story.html

Žižek, S. 2009a. *First as Tragedy then as Farce.* New York: Verso Books.

Žižek, S. 2009b. *In Defense of Lost Causes.* New York: Verso Books.

Žižek, S. 2012. *The Year of Dreaming Dangerously.* New York: Verso Books.

Zuboff, S. 2019. *The Age of Surveillance Capitalism: The Fight for a Human Future at the New Frontier of Power.* New York: Public Affairs.

4 The Geography of the 2020 Election's Presidential/Congressional Voting Gap

Adam S. Dohrenwend

In the weeks leading up to the 2020 U.S. federal election, Democrats had grown confident. Buoyed by an avalanche of favorable polling data and rosy forecasting by professional political prognosticators, the question regarding Democratic gains was not "if," but rather, "how much." Reputable outlets like FiveThirtyEight, *The Economist*, the University of Virginia's Center for Politics *Crystal Ball*, and the *Cook Political Report* suggested, among other things, an over 90% chance of Democratic victory in the presidential race, an expanded Democratic majority in the United States House of Representatives, and a Democratic United States Senate majority.

As polls closed, it quickly became clear that Democratic confidence had been misplaced. The results were far from assured. It took some days of counting in the crucial states of Pennsylvania, Georgia, and Arizona to confirm that Democratic presidential candidate Joe Biden had indeed won the election. Democrats lost an array of House seats (securing just four more than required for a 218 majority) and the Senate was extremely close, with two January 2021 runoff elections in Georgia left in the balance. Both Republican senatorial candidates Kelly Loeffler and David Perdue lost narrowly in the runoffs, giving the Democrats a majority with 50 seats by virtue of Vice President Kamala Harris' tie-breaking vote. While major forecasting outlets whiffed, the polling data underlying their predictions were fundamentally flawed.

Although President Biden earned over 81 million votes, more than any other presidential candidate in United States history and roughly seven million votes more than Donald Trump, scant to non-existent coattails left many down-ballot Democratic candidates short of victory, imperiling his legislative agenda. Of 148 competitive districts (those with a margin of victory for either or both the presidency and the House of Representatives), 95 (64%) featured races in which the Democratic candidate for Congress underperformed President Biden. Forty-seven featured races where this underperformance was greater than 5%. On the other hand, just 21 districts featured races where Republican candidates underperformed former President Donald Trump by more than 5%. These relative performance differences, while not as commonplace in prior decades, are especially important with partisan control of the House divided nearly evenly. While the adage "all politics is local"

DOI: 10.4324/9781003260837-4

may no longer carry the same weight in our current era of amplified electoral polarization and nationalization, this chapter underscores that candidates and their contexts still matter.

The Forecasts

FiveThirtyEight, founded by sports-statistics-turned-election-statistics guru Nate Silver, is among the most cited of political forecasters. The outlet's final "Deluxe" forecast proclaimed that "Democrats are *clearly favored to win the House*" – with 80% of model outcomes showing a 225 to 254 seat Democratic majority and an average seat count of 239.3 (a modest gain over the 235 seats won by Democrats in the 2018 midterm election). In 97 of 100 scenarios, Democrats were forecasted to win control (FiveThirtyEight 2020). In the outlet's "Lite" forecast, based solely on polling data, 98 of 100 scenarios forecasted continued Democratic control. Geoffrey Skelley, a FiveThirtyEight elections analyst, wrote hours before election day polls opened:

> Overall, the House contest appears to have significantly less drama than either the race for the presidency or the Senate, both of which are far more competitive. ... But perhaps the even bigger reason why Democrats are favored to keep control of the House – as well as maybe win the White House and even the Senate – is that the electoral environment looks quite good for their party. If we look at the polling average from our congressional generic ballot tracker, which includes all polls that ask respondents whether they plan to vote for the Democrat or Republican in their local congressional race, Democrats led by 7.3 percentage points.
>
> (Skelley 2020)

Yet the outcome was not nearly as rosy as the forecasters predicted. For example, in California, several seats in which Democrats were forecasted to win were lost. Democratic incumbents Gil Cisneros (District 39) and Harley Rouda (District 48) were rated with 74% and 68% chances of securing re-election, respectively. While Biden won in each district, both Congressmen lost (FiveThirtyEight 2020). In Florida, freshmen Democratic incumbents Debbie Mucarsel-Powell (District 26) and Donna Shalala (District 27) faced races rated as "Likely Democratic," with 82% and 81% chances of victory, respectively (FiveThirtyEight 2020). Both also lost re-election to their Republican challengers. Though Biden hemorrhaged support in heavily Latino areas of South Florida, as compared to the 2016 presidential tally, he still managed to win Shalala's Miami-based district. In Iowa, where three of four districts were held by the Democratic Party going into the election, Democrats Abby Finkenauer (IA-1), Rita Hart (IA-2), and Cindy Axne (IA-3) had 87%, 88%, and 84% forecasted chances of victory, respectively (FiveThirtyEight 2020). Only Axne won. Of the 50 most competitive races for the House, as projected by FiveThirtyEight, Democrats only won in ten.

The Economist's final forecast suggested that Democrats were "all but certain to keep their majority in the House of Representatives" – a greater than 99% chance of winning the required minimum of 218 seats (2020). Furthermore, *The Economist*'s forecast projected a 99% chance of the Democrats winning more than 222 seats – the number of seats they eventually (and actually) won (2020).

The Economist's forecasted median Democratic seat count stood at 244, of which 199 districts were deemed "Safe" (99% chance of Democratic victory), 34 districts were deemed "Very Likely Democratic" (85–99% chance of Democratic victory), and 10 districts were deemed likely Democratic (65–85% chance of Democratic victory). Twenty districts in total were deemed "Uncertain" (in which neither party had a greater than 65% chance of victory) – three of which were held by Democrats and 17 of which were held by Republicans (*The Economist* 2020). Ignoring these "Uncertain" districts, the forecast still provided Democrats with a 234-seat majority, one seat shy of their 2018 results.

Though the University of Virginia's Center for Politics' *Crystal Ball*, headed by Dr. Larry Sabato, correctly predicted 49 of 50 states in the presidential contest (only missing North Carolina), the outlet also predicted a net Democratic gain of ten seats in the House of Representatives – 243 seats in total (Sabato et al. 2020). The *Crystal Ball* correctly predicted 20 of 21 seats rated as "Likely Democratic" (only missing Shalala's race in FL-27), though many of the victorious candidates won by much narrower margins than expected. At the same time, Democrats lost 20 of 28 seats rated as "Leans Democratic." Nine "Leans Democratic" districts were held by Republicans going into election day, though the Democratic candidate won in only one (Bourdeaux, GA-7, where incumbent Republican Rob Woodall retired). On the other hand, Republicans swept all 40 districts rated as either "Leans Republican" or "Likely Republican." It is worth noting that the *Crystal Ball* does not use Toss-Up designations in their final forecasts. Reflecting on this inaccuracy, managing editor of the *Crystal Ball*, Kyle Kondik (2021), writes in his recently published book that much of the discrepancy between expectations and results can be attributed to Democrats' decreased support among Hispanic voters in many areas. Though polling data showed some warning signs for Democrats before election day, the extent of Republican gains among Hispanics did not become fully clear until returns were reported in South Florida. While Clinton won Miami-Dade County by 29% in 2016, Biden's 2020 margin decreased to just 7% (Kondik 2021). Far from an outlier, Democratic margins decreased across many of the nation's predominantly Hispanic communities, from Lawrence, Massachusetts to border areas in the Rio Grande Valley and in Southern California (Enten 2020). While Republicans were not able to win in most of these areas, their margin of loss to Democrats tended to decrease dramatically (Enten 2020).

The Cook Political Report, another longtime stalwart of election forecasting like the *Crystal Ball*, also missed the mark in its forecasting (Cook Political Report 2020). In contrast to the *Crystal Ball*, however, this outlet maintained its "Toss-Up" ratings in its final forecast. Of 18 districts rated as "Likely Democratic" and 19 districts rated as "Lean Democratic," Democrats lost in two (CA-39 and FL-27) and four (CA-48, FL-26, SC-01, and TX-23) districts, respectively. Though victorious in most of these districts, many Democratic candidates earned substantially lower vote tallies than expected. Of 26 districts rated as "Toss-Up," Republicans won all (9 of which were held by Democrats and 17 of which were held by Republicans – if one counts former Libertarian Congressman Justin Amash, a Trump critic who left the Republican party in 2019 and his seat open in 2020). Furthermore, Republicans won in all 15 districts rated as "Leans Republican" and 13 districts rated as "Likely Republican."

The Voting Gap: Ticket-Splitting and Voter Drop-Off

Voting gaps across the ballot are the product of two oft-studied phenomena: ticket-splitting and voter drop-off. Ticket-splitting occurs when voters select candidates from more than one party across their ballot – in this case, voting for one major party for the presidency and the other major party for Congress (Burden and Kimball 2002). Voter drop-off, often also called voter roll-off, is a phenomenon in which voters leave choices for some offices, usually down-ballot, blank (Pothier 1987). In this case, voter drop-off refers to selecting a candidate for the presidential race while leaving the congressional race blank. Though both practices demonstrate party disloyalty, ticket-splitting is a more pronounced measure.

Perhaps the most notable manifestation of these voting gaps is when they result in split or crossover districts – those districts that vote for one party for the presidency and another party for the House. Table 4.1 shows how the

Table 4.1 Crossover Districts in the US House, 1964–2020

Year	Total Districts	Crossover Districts	% of Total
1964	435	145	33.3
1968	435	139	32.0
1972	435	192	44.1
1976	435	124	28.5
1980	435	143	32.8
1984	435	190	43.7
1988	435	148	34.0
1992	435	100	23.0
1996	435	110	25.3
2000	435	86	19.8
2004	435	59	13.6
2008	435	83	19.1
2012	435	26	6.0
2016	435	35	8.0
2020	435	16	3.7

Source: Brookings Institution 2019; Coleman 2021.

number of crossover district results in each presidential election from 1964 to 2020. 2020's election featured crossover results in just 16, or 3.7%, of the 435 districts entitled to voting representation in Congress. Of the 148 congressional districts in which the presidential or congressional election was within a 15% margin, just over 10% featured split results.

As shown in Table 4.1, the share of crossover districts declined precipitously – a small fraction of the values seen in the past decades. During the decades where crossover districts made up as much as 44% of total districts, ticket-splitting was especially prominent. Many Southern Democrats began to support Republican candidates at the presidential level while still casting their ballots for down-ballot Democrats for Congress. Note that 1976 is a relative low point in terms of the number of crossover districts. That year, Jimmy Carter, a favorite son of the South, swept every former Confederate state besides Virginia – winning on the same ballot as many Democratic congressional candidates. It took decades for the Republican Party's "Southern Strategy" to solidify in down-ballot races.

Democrats still saw some significant successes down-ballot in rural and conservative districts through the 2004 and 2008 elections, with committee chairs like Ike Skelton (D-MO) and John Spratt (D-SC) consistently securing comfortable re-elections even as their districts voted for George W. Bush and John McCain. Most of this crossover support evaporated in 2010. Both lost, their districts representing two of 63 that Democrats lost that year. The share of crossover districts decreased by more than a third from 2008 to 2012, illustrating this massive loss of Democratic incumbents in the 2010 Republican wave election. As the parties and voting public have increasingly polarized, especially in the wake of the Trump candidacy and presidency, the share of crossover districts in 2020 is only about one-fifth the 2008 value.

There are two major lines of thought regarding ticket-splitting. The first, championed by researchers like Fiorina, is that voters intentionally split their tickets within a national frame of reference – especially so when partisan attitudes are highly polarized (Fiorina 1996). Called "balancing theory," this argument suggests that voters who split their tickets do so strategically to ideologically balance the results and following government. For example, while voting for Biden, a voter may also vote for a Republican candidate for Congress as a check on a hypothetical Democratic Biden administration. While perhaps attractive to national pundits, studies have shown that the voters least likely to split their tickets are those who perceive vast differences between the two major parties – undermining what Fiorina says is "probably the most important point" in his argument (Burden and Kimball 2002; Fiorina 1996; Kimball 2004). One need not look further than Table 4.1 to see that partisan polarization is not a primary driver of ticket-splitting.

The other major line of thought regarding ticket-splitting runs counter to balancing theory. Rather, differences across races on the same ballot are mostly borne out through personal and local attributes. This localized frame of reference stands in stark contrast with balancing theory's favoring of a national frame of reference. Burden and Kimball argue that "rather than a national mandate for bipartisan policymaking, ticket splitting has fundamentally local roots that are shaped indirectly the ideological positions of the national parties" and show "ticket-splitting is more common when the political parties converge toward the middle of the ideological spectrum [as it is] easier for voters to cross party lines when they do not have to travel far along the ideological spectrum" (2002, p. 38).

In the 2020 United States election, the two-party vote for the presidency totaled 155,485,078, while the two-party vote for Congress totaled 150,289,655. While third-party candidates did earn some votes beyond these totals, the primary factor at play in this difference is voter drop-off. It is typical for presidential contests to tally more votes than congressional contests – in part because of how high-profile presidential elections and their respective candidates are. While the average eligible voter may not know much, or anything, about the candidates running to represent them in Congress in any given election, it is much harder to ignore the ubiquitous visibility of a presidential contest. Voter drop-off is maximized in uncontested races – with many voters preferring to not cast a vote for a member of another party. While Black voters have been shown to drop-off down-ballot at higher rates than White voters, voter fatigue is another factor shown to contribute to ballot drop-off (Bullock and Dunn 1996; Vanderleeuw and Sowers 2007; Vanderleeuw and Utter 1993). This is a phenomenon in which voters, faced with a long ballot featuring many candidates for many offices, grow tired and leave some down-ballot races blank (Bullock and Dunn 1996).

One of the key challenges facing more detailed analysis of the voting gap is the age-old ecological fallacy, in which inferences are made about individual behavior based on inferences about the groups to which they belong (Burden and Kimball 1998; Burden and Kimball 2002; Gitelson and Richard 1983; King 1997). While researchers have followed approaches to more accurately gauge ticket-splitting and tease out the amount of voter drop-off without examining individual ballots, reliability in this context is questionable (Burden and Kimball 2002; Tam Cho and Gaines 2004). Furthermore, research on the specific nature of voter drop-off is relatively scant.

The Competitive Districts

Studying the margin gaps generated by ticket-splitting and voter drop-off at the district scale is key, as voters within their geographic boundaries determine the partisan composition of the House of Representative rather than the national electorate as a whole. As Burden and Kimball (2002, p. 35) state, "election outcomes are determined by aggregating votes within these electoral units." While elections occur in all House of Representatives constituencies, only a fraction are competitive – a phenomenon that has cemented itself over the course of the last century.

This chapter examines the voting gap specifically within these competitive districts using the official results as reported by the Office of the Clerk of the United States House of Representatives (Johnson 2017, 2021) and district-level presidential level results tabulated by the Daily Kos (2021). For the purposes of this chapter, a district is defined as competitive if either the presidential race or its respective congressional race in 2020 was decided by a margin of less than 15%. While 15% may seem like a significant margin to still count as competitive, one need not look further than the 2010 midterm election in which many numerous Democratic incumbents lost their seats two years after cruising to re-election in 2008. While the era of widespread ticket-splitting has "given way to strong party discipline among candidates and nationalized partisanship among voters," this tide of nationalization in our nearly-evenly divided Congress has made the relatively few persuadable voters especially pivotal (Kuriwaki 2021).

By limiting the analysis to these districts, I focus on the places where battles for national political mandates are won and lost. Furthermore, as uncompetitive districts often feature races that are only lightly contested, if at all, their inclusion would skew the results. For better or worse, it is a reality of the U.S. political system that a run for Congress takes serious funding and organizational muscle – and with limited resources, attention and resources are disproportionately provided to districts where they are most likely to make a difference: competitive districts.

According to the competitiveness standard outlined above, 148 districts, shown below in Figure 4.1, are defined as competitive. This is roughly one-third of the 435 districts that have voting representation in the House. According to the 2020 Census results, these districts average 62% non-Hispanic White (higher than the national figure of 59%) and 10% Black (lower than the national figure of 13%). Those identifying as Hispanic and/or Latino and Asian nearly mirror national figures – 18% and 7%, respectively. Few of the profiled districts are purely urban in character – as these districts generally feature lopsided Democratic victories across the ballot. Rather, these districts often consist of suburban territories of large metropolitan areas (like in the Los Angeles and Texas Triangle region) and rural swaths often interspersed with smaller metropolitan and micropolitan areas (like the Grand Rapids area and the many competitive districts of upstate New York).

In these competitive districts, Biden earned an average of 48.6% of the vote – slightly less than the 49.6% earned by Trump and just over 2.7% less than Biden's 51.3% nationally. By total votes across the competitive districts, Trump tallied just under 28.3 million votes to Biden's just over 27.6 million votes.

All 148 competitive districts featured a contested House race. Across these races, Republican candidates earned an average of 50.76% while Democratic candidates earned an average of 47.88%. On average, Republican candidates held an almost 12,000 average vote advantage – nearly triple the roughly 4,200 average vote advantage earned by Trump over Biden across these

Competitive Districts in the 2020 United States Presidential and Congressional Elections

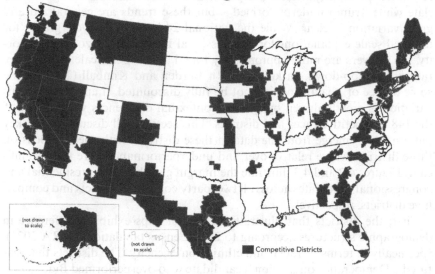

Figure 4.1 Competitive Congressional Districts, <15% Margin Victory for President and/or Congress, 2020.

Source: Author.

districts. By total votes for the House across these districts, Republican candidates tallied just over 28.4 million votes while Democratic candidates tallied just over 26.6 million votes.

Comparing the presidential race with congressional races shows that Biden earned an average of roughly 6,700 more votes per district over the Democratic congressional candidate, while Trump underperformed Republican congressional candidates by an average of 1,020 votes. Republican congressional candidates overperformed Trump's percent margin in 95 of 148 competitive districts, with an overperformance of greater than 5% or more in 47 of these. Just 21 of 148 districts featured the reverse, in which a Democratic congressional candidate overperformed Biden by 5% or more.

Contrary to conventional wisdom, in which, regardless of the results in percentages, candidates for president tend to earn more total votes than their down-ballot running mates, Trump actually earned over 150,000 fewer total votes across these districts than the combined vote totals for Republican congressional candidates. This fact highlights Trump's relative weakness among voters outside of his core base and further complicates his baseless claims of "election theft." Trump's electoral albatross, in the end, was himself. Furthermore, these figures indicate the limited salience of coattails on either side of the partisan divide in the 2020 House elections – Biden overperformed Democratic congressional candidates while Trump underperformed Republican congressional candidates.

Patterns of Congressional Overperformance and Underperformance

As stated previously, Biden overperformed the Democratic congressional slate while Trump underperformed – but these trends are generalized and great variation is clear. While the national 24-hour news cycle grasps for national scale explanations for congressional Democrats' razor-thin majority, the answers are more appropriately found at the local scale: bottom-up, rather than top-down. In keeping with Burden and Kimball (2002), while some degree of "balancing" cannot be fully discounted, it appears that this variation is greatly driven by specific contextual differences within each of the 148 competitive races and districts. This section will discuss several key patterns that emerge from the data in these districts, with specific focus on those districts where relative over and underperformance were most significant. Figures 4.2 and 4.3 illustrate the margin gaps between presidential and congressional candidates across all two-party contested districts and competitive districts, respectively.

First, the districts that illustrate the largest gaps exhibit a difference in demographic structures according to 2020 Census tabulations (DRA 2021), specifically in terms of racial and ethnic homogeneity. The districts that featured a Democratic congressional candidate who overperformed Biden's percent margin by 10% or more in their district tend to feature more homogenous populations. In these six districts, the average percentage share of the population by the largest racial or ethnic demographic group is 82%. Inclusion of the seventh-strongest showing by a Democratic congressional candidate relative to Biden (by Filemon Vela, TX-34 anchored in the Rio Grande Valley and 85% Hispanic or Latino), outperforming by just under 10%, would increase the average figure to 83%. The lowest value is found in the Rio Grande Valley's TX-28, represented by longtime conservative Democrat Henry Cuellar, where 77% of the population identify as Hispanic or Latino. The highest value is seen in ME-2, represented by Jared Golden, where 91% of the population identifies as White.

The districts on the opposite side of the spectrum, where 16 Republican congressional candidates outperformed Trump by 10% of more in their district, this figure is 68%. Unlike the parallel group of overperformers on the Democratic side, all of these candidates won their races. While several of these districts are relatively homogenous, like those won by Ohio Republicans Gonzalez, Stivers, and Joyce, many have high degrees of racial and ethnic heterogeneity – like CA-25 (Garcia-R), CA-39 (Kim, R), and TX-2 (Crenshaw-R). None of these districts have a single racial or ethnic group reaching a majority in their respective populations and feature higher than average shares of those identifying as Hispanic and Latino or Asian. While this claim is based on small sample sizes, a clear difference is apparent and could warrant future research. Although areas with higher racial diversity are more amenable to Democrats, party disloyalty is more common down-ballot than in presidential races – potentially setting the stage for overperformance by Republican congressional candidates in these districts (Bader 2020; Burden and Kimball 2002).

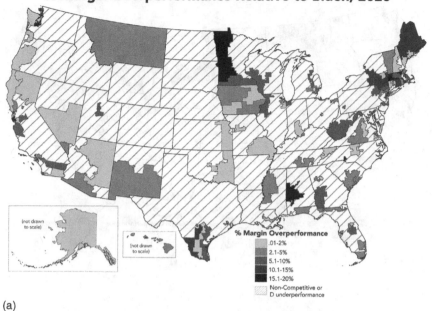

(a)

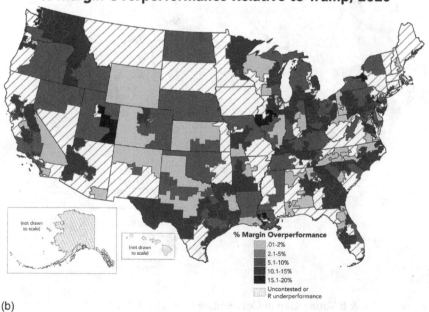

(b)

Figures 4.2 a & b Voting-Gap in All Two-Party Contested Districts, 2020.
Source: Author.

Democratic Congressional Candidate % Margin Overperformance Relative to Biden in Competitive Districts, 2020

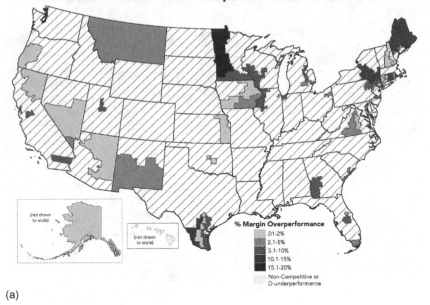

(a)

Republican Congressional Candidate % Margin Overperformance Relative to Trump in Competitive Districts, 2020

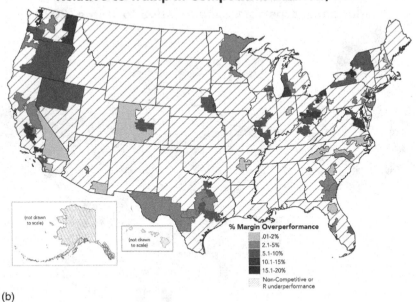

(b)

Figures 4.3 a & b Voting-Gap in Competitive Districts, 2020.
Source: Author.

Competitive districts anchored, at least in part, in the inner suburbs of large metropolitan areas, as well as those dominated by a single, whole or nearly-whole smaller metropolitan area, were more likely to see higher levels of Republican congressional candidate overperformance compared to Donald Trump. The districts of Republican incumbent Congressmen John Katko (NY-24, centered on the Syracuse metropolitan area) and Brian Fitzpatrick (PA-01, almost wholly contained in Bucks county, a relatively wealthy Philadelphia suburb) had the highest and second-highest overperformance values of all 148 profiled competitive districts. Both won re-election as their districts voted for Biden – surpassing Trump performance in their districts by roughly 19% each.

A similar situation, though to a lesser degree, is seen in NE-02 (dominated by Omaha and its inner suburbs), where incumbent Republican Don Bacon won re-election while Trump lost his district (and a coveted electoral under the state's split allocation method). Bacon won by about 4.5% and Trump lost by about 6.5% – a net overperformance of about 11%. The undervote for Congress in this district, a marker of voter roll-off, was less than 800 (among the lowest here of all competitive districts).

Though Bacon has been more conservative than Katko and Fitzpatrick, he is known to criticize Trump's rhetoric. Attracting ticket-splitters, Bacon survived his challenge from progressive Kara Eastman, who had beaten a more moderate, establishment-backed candidate in the Democratic primary. Bacon defeated moderate incumbent Democrat Brad Ashford in 2016. In Ashford's attempt at a 2018 rematch, he narrowly lost in the Democratic primary to Eastman, who then lost to Bacon. Running again in 2020, Eastman beat Ann Ashford (Brad Ashford's wife) in the Democratic primary by a much larger margin. After losing the primary, Ann Ashford declined to endorse Eastman while Brad Ashford endorsed his former rival Bacon (Morton 2020).

Katko and Fitzpatrick are seen as moderate Republicans. Both have been skeptical of Donald Trump and his leadership over their party, joining a handful of their Republican colleagues in voting to impeach him in the aftermath of the January 6 U.S. Capitol riot, and are known to buck their party on some key policy issues. Young Kim (CA-39), another Republican overperformer, narrowly won a rematch against freshman Democrat Gil Cisneros by just over 1% in a plurality-Asian district anchored in Southern California's suburban Orange County, long a Republican stronghold.

An immigrant from South Korea, Kim joined Michelle Steel (CA-48) in victory as the first Korean-American representatives from California in over 20 years and the first female Korean-American representatives in U.S. history alongside Marilyn Strickland (WA-10). Biden won Kim's district by just over 10%, making her one of nine House Republicans to hold a Biden district. In contrast, Steel, who also now represents a Biden district based in Orange County and long previously held by Republicans (CA-48), staked out more socially conservative positions and had adopted her husband's non-Korean surname. She earned significantly less crossover support than Kim, but still managed to defeat freshman Democrat Harley Rouda. Biden won the district by a narrow 1.5%, while Steel won by 2.1%, a margin overperformance of 3.6%.

Kim had the highest overperformance value of any non-incumbent candidate in a competitive congressional district other than David Valadao (CA-21), that district's former congressman. Valadao faced a re-match against T.J. Cox, a freshman Democrat who unseated Valadao in 2018. Valadao won by just under 1%. Aided by scandals involving Cox's past financial dealings, Valadao's coolness towards Trump's immigration policies and past work on irrigation also likely played a role his comeback. The district is 75% Hispanic or Latino and is based in California's San Joaquin Valley, a major agricultural region dependent on migrant labor and facing water issues (Irby 2020).

Similar moderation on immigration and environmental issues may have also played a role in Republican María Elvira Salazar's (FL-27) 2.75% rematch victory over freshman Democrat Donna Shalala in a district Biden won by 3.2%. Salazar, a retired five-time Emmy Award-winning television journalist for Univision, had deep cultural ties to the district. The daughter of Cuban exiles, Salazar was born in Miami's Little Havana (contained within this 70% Hispanic/Latino district) and faced a Democratic opponent who did not speak Spanish and was serving at age 79 as the second-oldest freshman representative in U.S. history (Cochrane 2018; Smiley and Tavel 2018).

Incumbency is usually among the strongest advantages a candidate can have in a congressional race and longtime representatives with a long-tended positive reputation in their district's communities often benefit the most. Incumbency can foster or provide a variety of benefits, including higher name recognition, already-established constituent services, the franking privilege, the ability to take an active policy role in popular local issues through oversight and legislation, perhaps most importantly, often advantages in campaign fundraising. Furthermore, since they have established congressional positions, incumbents have more ease in differentiating themselves from party orthodoxy – often key in generating high overperformance values relative to their party's presidential candidate. This is especially useful in districts that have seen significant partisan shifts in their election results during the Trump era.

Areas where partisanship has shifted in recent years are hallmarks of many of the districts on either side of the overperformance spectrum – illustrating that, like in the South previously, shifts in partisan preferences for Congress tend to lag shifts in partisan presidential preferences when constituencies realign. Dan Kildee (MI-05) replaced his uncle, longtime representative Dale Kildee, in 2013 and swept his little-known 2020 challenger by almost 13%. At the same time Trump came within 4% of winning the district. While Kildee earned almost 7,400 more votes in 2020 than Biden did in the district, Republican candidate Tim Kelly earned almost 25,000 fewer votes than Trump.

Facing off against a highly-visible incumbent whose family has represented the area in Congress since 1977, Republicans chose not to prioritize Kildee's seat and a fair number of Trump voters either voted for Kildee's re-election or left that portion of the ballot blank. Had Kildee not run in 2020, the race for his seat would surely have been more competitive. On the Republican side, Congressman Fred Upton (MI-06) provides a similar example. In Congress since 1987, Upton is an affable Representative who, like most moderate Republican Representatives, kept Trump at arm's length and on occasion even condemned his rhetoric. Though facing legitimate opposition in 2020, Upton won the district by over 15% – overperforming Trump's 4.5% by more than 11%.

Collin Peterson (MN-07) and Jared Golden (ME-02), the highest and third-highest Democratic overperformers of Biden, respectively, were the only two Democrats to vote against one or more articles of impeachment during Trump's first trial in 2019. This is just one of a variety of high-profile issues on which Peterson and Golden had broken with party leadership. Peterson voted against both articles. Peterson's chairmanship of the House Agriculture Committee is also of note, as MN-07 is a center of sugar beet production. Peterson had the highest overperformance of any Democratic candidate in a competitive district relative to Biden (almost 16%), but still lost his bid for a 16th term by 13.5%. Trump won the district by almost 29%, a much larger win than Mitt Romney had in the district, but Republican challenger Michelle Fischbach earned almost 41,000 fewer votes than Trump.

Like Peterson, Golden, too, represented a very rural district (the largest by area east of the Mississippi River) that had swung sharply right during the Trump era. Both districts Golden voted for Article #1 but against Article #2 in 2019, achieving the same functional result (impeaching Trump) but allowing himself distance from an effort unpopular in his district. Staking out progressive positions on public financing of elections, union rights, and the minimum wage, coupled with more conservative positions on other issues like Trump's first impeachment trial and gun rights, Golden was able to appeal to some blue-collar and populist-leaning rural Trump voters with a recent history of voting for Democrats at the presidential level.

Golden's overperformance was the third-best among Democratic Congressional candidates – beating Biden's 7.5% loss by 13.5%, in a 6% win overall. Like Katko and Fitzpatrick, Golden shows that breaking with party orthodoxy at least some of the time can pay electoral dividends. One of just seven Democrats elected in a Trump-won district in 2020, Golden has maintained distance from party leadership positions, with an article in *Politico* quoting him that while "not a centrist … no amount of pressure" can change his mind on a vote (Ferris 2021). As noted in the article's headline, with Peterson's defeat, Golden now represents "Democrats' Trumpiest district" (Ferris 2021).

In addition to MN-07 and ME-02, the vast majority of districts featuring large overperformances by a Congressional candidate over their respective party's presidential nominee feature incumbents seeking re-election. Freshman Democrat Anthony Brindisi (NY-22) unseated Claudia Tenney in 2018 but faced a 2020 rematch in this central New York district. Brindisi heralded his success in passing the SPOONSS Act, an initiative his predecessor introduced but failed to enact due to resistance from Republican leadership. The initiative, signed into law in 2019 by Trump, requires the United States Department of Defense to buy forks, knives, and spoons produced in the United States (Weiner 2019). It is not a coincidence that the district is home to Sherrill Manufacturing, the country's only manufacturer that sources all its eating utensils domestically (Weiner 2019).

In another nod to district-specific issues that cut across party lines, Brindisi frequently highlighted his efforts to combat poor service and unfair practices by Spectrum, a major cable provider in Upstate New York and frequent target of public ire (Eames 2020; Weiner 2020). It may come as a surprise that oversight of Spectrum had become a marquee issue in the race, with both candidates frequently raising the issue (Howe 2020).

Though a more reliable vote for leadership than Golden and Peterson, Brindisi, like many overperforming congressional candidates, sought to differentiate himself from a generic Democrat by giving special attention to and taking populist positions on local issues. By being an incumbent, Brindisi was able to back up his rhetoric with concrete official actions. While Brindisi lost re-election after months of adjudication by 109 votes (of over 325,000 total cast), he had the fourth-highest margin outperformance of Biden among Democratic congressional candidates (11.5%). As Trump won the district by 11.5% (a much higher margin than Romney had won by in 2012), had Brindisi won, he would take Golden's designation as the representative of "Democrat's Trumpiest District."

Though it may be tempting to draw the conclusion that ideological moderation is the key to congressional candidate overperformance, Congressional Progressive Caucus (CPC) member and Matt Cartwright (PA-8) provides a foil. First elected in 2012 after defeating moderate ten-term incumbent Tim Holden in that year's Democratic primary from the ideological left, Cartwright represents the most Republican-leaning district of all CPC members. Anchored by Scranton and Wilkes-Barre and known for its populist leanings, the area has trended rightward during the Trump era. While Trump won in PA-8 by almost 4.5%, Cartwright won by 3.5%, outperforming Joe Biden by about 8% in his boyhood turf. Cartwright's margin outperformance was the 10th largest of any Democratic congressional candidate in a competitive district.

By total votes, Cartwright earned over 9,700 more votes than Biden while Republican challenger and former Trump Export–Import Bank appointee Jim Bognet earned over 18,600 fewer votes than Trump. There were almost 13,000 two-party undervotes for Congress in PA-8 – nearly double the average in competitive districts. While Trump won here, nearly a sixth of his voters here either split their ticket for Cartwright or left this portion of their ballot blank.

Conor Lamb (PA-17) illustrates the other side of this issue in Pittsburgh's largely White, northwestern suburbs. Lamb, a Problem Solvers Caucus member, has carefully manicured his image as a "normal Democrat" who wins competitive elections while now seeking the 2022 Democratic nomination for

U.S. Senate in Pennsylvania. Seeking to contrast himself with progressive populist and fellow frontrunner Lieutenant Governor John Fetterman, Lamb sees his record and three congressional wins since 2018 as a selling point in his current Democratic primary race for the United States Senate (Lamb 2021; Tamari 2021). Lamb narrowly won a 2018 special election (by 0.4%) in Pennsylvania's then-18th district to succeed Tim Murphy, a Republican, who resigned after details of his extramarital affair became public. Two years earlier, Trump won this district by just under 20%.

Under new congressional district lines adopted by the commonwealth's Supreme Court during his special election campaign, Lamb opted to run in the new PA-17 that fall – where Trump would have narrowly won in 2016. Though Lamb sailed to a double-digit victory that fall in his new district, he just narrowly held on against Republican nominee Sean Parnell in 2020 – by 2.2%. At the same time, Biden won the district by 2.7%. Lamb earned almost 750 more votes than Biden did here, but Parnell earned almost 2,700 more votes than Trump.

Lamb's notable performances in areas Trump won in 2016 evaporated in a high-turnout presidential election. Of note, Parnell, though facing domestic abuse allegations, was viewed by many as the Republican primary frontrunner in the race, until he had to withdraw (Medina 2021; Tamari 2021). As suggested earlier, especially with the potential for 2022 Republican gains in the House, branding as a "normal" member of one's party is not necessarily the performance driver that Lamb and his supporters hope for. Rather, many of those candidates who outperform relative to the topline partisanship of their districts actively seek to distinguish themselves – making a play for ticket-splitters and emphasizing local issues rather than their own partisan typicalness.

In many districts, relying on the standard "national" coalition mounted by a candidate's party, is often insufficient to win. To be successful, candidates in these districts must build district-wide coalitions that often look quite different from the national coalition. Simply put, if the national party struggles to win a district in presidential races (the most nationalized of all), tightly gripping party orthodoxy is not likely to be the best strategy to win. If it were, the party would not see such difficulty in the presidential contest in the first place.

Conclusions

While much attention is given to gaps between votes for President and Congress in relation to divided government, the results from 2020 in which a Democratic President was elected alongside a diminished Democratic majority in the House, provide an interesting dynamic for the 117th United States Congress. While Donald Trump still holds enormous sway over congressional Republicans, *most of them actually did better than Trump in 2020*, including many from competitive districts. A variety of high-profile votes in the early months of the 117th Congress demonstrate the apparent lack of fealty felt by some of these congressional Republicans on issues directly related to Trump and his political survival.

First is the set of votes on the objection to counting the electoral votes of Arizona and Pennsylvania in the certification of the 2020 presidential election. While the majority of House Republicans voted to sustain the objection – 121 for the objection in Arizona and 138 for the objection in Pennsylvania – just two of the 16 significant (by more than 10%) margin outperformers of Trump in competitive districts (Garcia, CA-25 and Wittman, VA-01) voted to sustain one or both objections. Though Valadao did not vote, as his swearing-in was delayed by a few days due to COVID-19 infection, the other 13 voted against both objections.

Trump's second impeachment trial a few weeks later shows another example. Though just ten House Republicans voted to impeach Donald Trump over the January 6 attack on the Capitol and the event's prior machinations, nine of them outperformed Trump in 2020 – with underperformer Liz Cheney (WY) as the lone exception. While not all these ten represent districts are defined as competitive, eight of these representatives, including Katko (NY-22), Kinzinger (IL-16), Gonzalez (OH-16), Herrera-Beutler (WA-3), Newhouse (WA-4), Upton (MI-6), Valadao (CA-21), and Newhouse (WA-4), substantially outperformed Donald Trump.

Another example of this dynamic, with a look by candidates to 2022, is seen in the November 2021 House vote on the Infrastructure Investment and Jobs Act – the bipartisan infrastructure deal developed in the Senate earlier in the summer. Thirteen Republicans, much to the chagrin of their congressional leadership and Donald Trump, voted for the bill. All but two, Kinzinger (IL-16) and McKinley (WV-01), represent competitive districts. Unlike in the votes previously discussed, this Republican support for the bill was decisive, allowing it to be sent to President Biden's desk over the objections of six progressive House Democrats. While not all overperformed Trump in 2020, several did by a substantial amount, like Bacon (NE-02), Fitzpatrick (PA-01), Gonzalez (OH-16), Katko (NY-22), Smith (NJ-04), and Upton (MI-06). Had Democrats not underperformed Biden in 2020's battle for Congress and won a majority as large or larger than that elected in 2018, these Republican votes would likely have been inconsequential.

In conclusion, I look to the future. Nicole Malliotakis (NY-11), a freshman Republican who defeated incumbent Democrat Max Rose in 2020 but underperformed Trump by several percent, provides a cogent example of the drive to win in politically unfriendly territory. She voted for the bipartisan infrastructure deal but has expressed staunch opposition to the related Build Back Better Act – the larger reconciliation package still working its way through Congress. Though her current district voted for Trump by over 10%, Malliotakis's eyes are on the ongoing redistricting process in New York. Eager to wield their legislative supermajorities, Democrats here are considering an aggressive gerrymander producing a map in which 23 of 26 congressional districts voted for Biden (Fandos and Ashford 2021). Malliotakis's Staten Island and south Brooklyn district is directly in state legislators' crosshairs, easily made into a Biden district by slightly adjusting its territories in Brooklyn.

Looking to the future, Malliotakis is splitting her votes – much like she hopes Biden voters might in her future races in a hypothetical new Democratic-leaning district. With redistricting looming, her future prospects likely rest on becoming an overperformer of up-ticket Republicans. This is not a national-ized strategy, but, rather, one that illustrates the salience of personal and local factors in driving party disloyalty when voting for Congress. While races for Congress are certainly more nationalized than in past decades, and there is earlier precedence for this going back to the late 1800s, nationaliza-tion is not absolute. Contrary to narratives of nationalization spouted by cable news pundits and underscored by the current knife's-edge partisan division in Congress, candidates and their contexts still matter.

•

References

Bader, M. 2020. Can racial diversity swing competitive Congressional elections? *Contexts* 19(2):68–70.

Brookings Institution. 2019. *Vital Statistics on Congress*. Washington, DC: Brookings Institution. https://www.brookings.edu/VitalStats

Bullock, C., III and R. Dunn. 1996. Election roll-off: A test of three explanations. *Urban Affairs Review* 32(1):71–86.

Burden, B. and D. Kimball. 1998. A new approach to the study of ticket splitting. *American Political Science Review* 92(3):533–544.

Burden, B. and D. Kimball. 2002. *Why Americans Split Their Tickets*. Ann Arbor: University of Michigan Press.

Cochrane, E. 2018. Too old to be a freshman in Congress? Donna Shalala doesn't care. *New York Times* (Dec. 30). https://www.nytimes.com/2018/12/30/us/politics/donna-shalala-congress.html

Coleman, J. 2021. 2020's crossover districts. *Sabato's Crystal Ball*. University of Virginia Center for Politics. https://centerforpolitics.org/crystalball/articles/2020s-crossover-districts/

Cook Political Report. 2020. 2020 House race ratings. Oct. 21. https://cookpolitical.com/ratings/house-race-ratings/230616

Daily Kos. 2021. Daily Kos elections' statewide election results by congressional and legislative districts. March 23. https://www.dailykos.com/stories/2013/07/09/1220127/-Daily-Kos-Elections-2012-election-results-by-congressional-and-legislative-districts

DRA. 2021. Congressional district demographics. Dave's Redistricting App. https://davesredistricting.org/

Eames, S. 2020. Rep. Brindisi says Spectrum is scamming customers. *The Daily Star* (Jan. 9). https://www.thedailystar.com/news/local_news/rep-brindisi-says-spectrum-is-scamming-customers/article_4b95194a-0e3c-5c40-ae2f-b9e1bcbac188.html

The Economist. 2020. Forecast the US elections: House. (Nov. 2). https://projects.economist.com/us-2020-forecast/house

Enten, H. 2020. Trump made big in-roads in Hispanic areas across the nation. *CNN* (Dec. 12). https://www.cnn.com/2020/12/12/politics/trump-hispanic-vote/index.html

Fandos, N. and G. Ashford. 2021. New York will soon lose 1 House seat. The G.O.P. might lose 5. *New York Times* (Sept. 14). https://www.nytimes.com/2021/09/14/nyregion/congress-redistricting-ny.html

Ferris, S. 2021. The former Marine who holds Democrats' Trumpiest district. *Politico* (May 11). https://www.politico.com/news/2021/05/11/jared-golden-maine-democrats-trump-486559

Fiorina, M. 1996. *Divided Government.* 2nd ed. Needham Heights, MA: Allyn and Bacon.

FiveThirtyEight. 2020. 2020 House forecast. (Nov. 2). https://projects.fivethirtyeight.com/2020-election-forecast/house/

Gitelson, A. and P. Richard. 1983. Ticket-splitting: Aggregate measures vs. actual ballots. *Western Political Quarterly* 36(3):410–419.

Howe, S. 2020. NY-22: Brindisi, Tenney on police reform, spectrum, more. *Observer-Dispatch* (Sept. 14). https://www.uticaod.com/story/news/2020/09/14/ny-22-brindisi-tenney-on-police-reform-spectrum-more/42617353/

Irby, K. 2020. Did GOP's Valadao work with Democrats on healthcare, water and immigration? *Fresno Bee* (Aug. 19). https://www.fresnobee.com/news/local/article245043320.html

Johnson, C. 2017. Statistics of the Presidential and Congressional Election of November 8, 2016. Office of the Clerk of the U.S. House of Representatives (Feb. 22). https://history.house.gov/Institution/Election-Statistics/Election-Statistics/

Johnson, C. 2021. Statistics of the Presidential and Congressional Election of November 3, 2020. Office of the Clerk of the U.S. House of Representatives (Feb. 28). https://history.house.gov/Institution/Election-Statistics/Election-Statistics/

Kimball, D. 2004. A decline in ticket splitting and the increasing salience of party labels. In *Models of Voting in Presidential Elections: The 2000 Elections*, H. Weisberg and C. Wilcox (eds.) Palo Alto: Stanford University Press.

King, G. 1997. *A Solution to the Ecological Inference Problem.* Princeton: Princeton University Press.

Kondik, K. 2021. *The Long Red Thread.* Athens, OH: Ohio University Press.

Kuriwaki, S. 2021. The swing voter paradox: electoral politics in a nationalized era. PhD dissertation. Harvard University Graduate School of Arts and Sciences.

Lamb, C. 2021. Twitter post. (Nov. 7). https://twitter.com/ConorLambPA/status/1457553657898184705

Medina, J. 2021. Sean Parnell suspends G.O.P. Senate bid in Pennsylvania. *New York Times* (Nov. 22). https://www.nytimes.com/2021/11/22/us/politics/sean-parnell-suspends-pennsylvania-senate.html

Morton, J. 2020. Republican Don Bacon wins endorsement of former rival, Democrat Brad Ashford. *Omaha World-Journal* (Oct. 7). https://omaha.com/news/local/govt-and-politics/republican-don-bacon-wins-endorsement-of-former-rival-democrat-brad-ashford/article_7ec2bd40-e6e3-5dfb-988f-9ec081e07df2.html

Pothier, J. 1987. Drop-off, the vanishing voters in on-year elections, and the incumbency advantage. *American Politics Quarterly* 15(1):123–146.

Sabato, L., K. Kondik, and J. Coleman. 2020. *Final Ratings for the 2020 Election.* Sabato's Crystal Ball, University of Virginia Center for Politics (Nov. 2). https://centerforpolitics.org/crystalball/articles/21320/

Skelley, G. 2020. Final forecast: Democrats are clear favorites to maintain control of the House. *FiveThirtyEight* (Nov. 3). https://fivethirtyeight.com/features/final-2020-house-forecast/

Smiley, D. and J. Tavel. 2018. The Shalala conundrum: Wooing Hispanic voters when you don't speak the language. *Miami Herald* (Oct. 16). https://www.miamiherald.com/news/politics-government/election/article220126680.html

Tam Cho, W.K. and B. Gaines. 2004. The limits of ecological inference: The case of split-ticket voting. *American Journal of Political Science* 48(1):152–171.

Tamari, J. 2021. Pa. Senate candidates filed their latest fund-raising reports. Here's what they show. *Philadelphia Inquirer* (Oct. 16). https://www.inquirer.com/news/senate-pennsylvania-fundraising-candidates-20211016.html

Vanderleeuw, J. and T. Sowers. 2007. Race, roll-off, and racial transition: The influence of political change on racial group voter roll-off in urban elections. *Social Science Quarterly* 88(4):937–952.

Vanderleeuw, J. and G. Utter. 1993. Voter roll-off and the electoral context: A test of two theses. *Social Science Quarterly* 74(3):664–673.

Weiner, M. 2019. House OKs Anthony Brindisi's bill to boost America's last flatware maker. *Syracuse.com* (July 15). https://www.syracuse.com/politics/2019/07/house-oks-anthony-brindisis-bill-to-boost-americas-last-flatware-maker.html

Weiner, M. 2020. Rep. Anthony Brindisi asks NY to investigate Spectrum's weekend outage. *Syracuse.com* (Feb. 10). https://www.syracuse.com/news/2020/02/rep-anthony-brindisi-asks-ny-to-investigate-spectrums-weekend-outage.html

5 Contesting Control of the Senate
The Georgia Senate Elections, 2020–2021

Fred M. Shelley

In 2020, the state of Georgia proved critical to the outcomes of the elections for the presidency and the Senate nationwide. Georgia was one of five states that gave its electoral votes to Republican Donald Trump in 2016 but switched to his Democratic challenger Joe Biden four years later. In January 2021, Democratic challengers Jon Ossoff and Raphael Warnock defeated Republican incumbents David Perdue and Kelly Loeffler in runoff elections for Georgia's two seats in the Senate. Their success gave the Democrats majority control of the Senate and therefore control of both houses of Congress. In this chapter, I examine geographic dimensions of the outcomes of these Senate elections and compare these results with previous elections in order to illustrate the changing demographics and political geography of Georgia.

Georgia's Presidential Elections in Historical and Contemporary Perspective

Georgia's electoral history parallels that of many of its Southern neighbors. After Reconstruction in the 1870s, conservative white Democrats gained control of the state legislature. Over the ensuing decades, "Jim Crow" laws which disenfranchised most Blacks and some poor whites were enacted, reinforcing the dominance of Georgia politics by state's conservative elites. Hence, the Democratic Party came to dominate Georgia's politics at the state and national levels. Democratic nominees won Georgia's electoral votes in every election from 1876 through 1960.

Between 1917 and 1962, Georgia's statewide primary elections were decided by a county unit electoral system (Buchanan 2005). Under this procedure, Georgia's 159 counties were divided into three categories: "urban", "town", and "rural". The eight "urban" counties were awarded six unit votes each, with the 30 "town" counties awarded four unit votes each and the 121 "rural" counties awarded two votes each. Although this system was supposedly modeled after the Electoral College, it had the practical effect of privileging rural counties at the expense of urban counties. The "rural" counties had a combined total of 242 unit votes, whereas the others had a combined total of 168 unit votes. This disparity increased as urban populations

DOI: 10.4324/9781003260837-5

increased relative to rural populations, especially after World War II. The system reinforced conservative elite control over Georgia's politics until it was declared unconstitutional by the U.S. Supreme Court in *Gray* v. *Sanders* (372 U.S. 368) in 1963.

The disenfranchisement of most Blacks in Georgia continued until Congress enacted the Voting Rights Act of 1965. A year earlier, the primarily white Georgia electorate delivered a majority to Republican Barry Goldwater, who had voted against the Civil Rights Act earlier that year. Although the Voting Rights Act had authorized the federal government to enforce the rights of Blacks to vote, neighboring Alabama's George Wallace, who was a strong proponent of segregation and opponent of the Civil Rights move-ment, carried Georgia in 1968. Significantly, however, Wallace's margin of victory in Georgia was considerably smaller than his margins of victory in neighboring Southern states.

From the 1970s onward, the Republican Party became more competitive in Georgia. Republican incumbent Richard Nixon carried Georgia in 1972. In 1976, Georgia's former governor Jimmy Carter carried his native state along with every other state in the former Confederacy except for Virginia. In winning the 1976 election, Carter became the first president from the Deep South since the Civil War. Carter carried Georgia again in 1980 in his unsuccessful re-election bid against Republican Ronald Reagan, although Carter lost the remaining Southern states. Subsequently, the Republicans became the dominant party at the presidential level in Georgia. Arkansas native Bill Clinton carried Georgia narrowly in 1992, but he lost Georgia four years later. Republican nominees won Georgia's electoral votes in the next six elections.

In 2016, Trump won Georgia by a margin of 211,137 popular votes over his Democratic opponent, Hillary Clinton. In 2020, however, Biden became the first Democrat since 1992 to carry Georgia when he won the state's elec-toral votes by a margin of 11,779 popular votes over Republican incumbent Trump. Biden won Fulton County, which contains the city of Atlanta, by more than 240,000 popular votes. He also carried the suburban counties sur-rounding Atlanta, winning considerably higher percentages of the vote in these counties than had Hillary Clinton four years earlier. Comparison of results at the precinct level between 2016 and 2020 illustrates the degree to which voting shifts in Atlanta's suburbs were key to his statewide victory, as illustrated in Figure 5.1 (Cohn et al. 2020). Biden's majorities in the metro-politan Atlanta area were sufficient to overcome Trump's majorities in other parts of the state. Because large metropolitan areas such as the Atlanta area are growing at faster rates than rural areas, Biden's victory in Georgia reflects these demographic changes and could presage similar shifts in other Southern states (Slodysko 2021).

Trump and his supporters contested the outcome of the election in Georgia fiercely. A week after the election, Georgia's Republican Secretary of State Brad Raffensberger ordered a statewide recount of the ballots. The recount resulted in a net increase of 1,274 votes for Trump, nowhere near the more

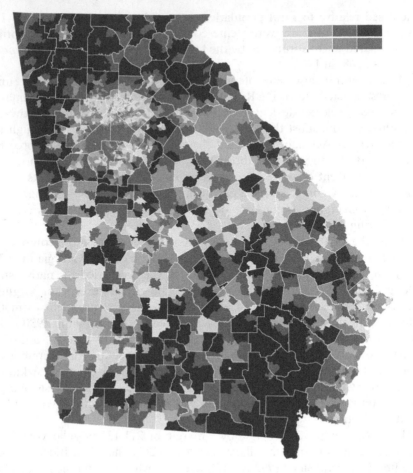

Figure 5.1 Precinct-level changes in presidential vote between 2016 and 2020 in Georgia.

Source: https://www.nytimes.com/interactive/2020/11/17/upshot/georgia-precinct-shift-suburbs. html.

than 11,000 votes that Trump would have needed to overtake Biden. The Trump campaign attempted to delay the certification of the Georgia results further, but this motion was denied by a Federal judge (Fowler 2020). Raffensberger certified Biden as having won Georgia's electoral votes on December 7, and its electoral votes were cast for Biden a week later. However, Trump continued to contest the results and in early January he telephoned Raffensberger and asked him to "find" enough additional votes needed for him to win the state, although the results had already been certified (Gardner 2021). For months after Biden took office, Trump and many of his supporters continued to claim that the election in Georgia and in other states had been stolen by Biden and the Democrats.

The 2020 Senate Elections in Georgia

While voting in the presidential election, Georgia voters also cast ballots in elections for both of the state's seats in the United States Senate. A combination of circumstances resulted in these elections determining party control of the Senate, thereby putting the state in the national political spotlight for several weeks after the presidential election.

Although the Republicans won the electoral votes of Georgia in all but one presidential election between 1984 and 2016, they were not nearly as successful in U.S. Senate races. During that period, four Democrats – Sam Nunn, Wyche Fowler, Max Cleland, and Zell Miller – won Senate elections from Georgia as did four Republicans – Paul Coverdell, Saxby Chambliss, Johnny Isakson, and Perdue. Perdue was first elected to his seat in 2014, replacing Chambliss, who had retired from the Senate after having served two terms.

Georgia's other seat became open after Isakson resigned from the Senate on December 31, 2019, because of ill health. Isakson had been elected to the Senate originally in 2004 and was re-elected to his third term in 2016. According to Georgia law, when a vacancy occurs during a Senate term, the governor appoints a replacement to fill the vacancy temporarily. A special election is then held to fill the remaining years of the term – in this case, for the remaining two years of Isakson's term that had begun after the 2016 election. In accordance with this procedure, Republican Governor Brian Kemp appointed Kelly Loeffler to fill the seat.

Together, the two Senate elections in Georgia became critical because of the partisan division of the Senate nationwide and because Georgia law requires the winner of a statewide election to win a majority of votes cast. After the votes were counted throughout the country following Election Day, the Republicans had won 50 of the 100 seats in the Senate compared to 46 for the Democrats. (Independents Senators Angus King of Maine and Bernie Sanders of Vermont caucus with the Democrats, giving them an effective total of 48 seats.) However, the two Georgia Senate races remained undecided after Election Day.

According to Georgia law, if no candidate for an office gets a majority of the vote on Election Day, then a runoff election is held subsequently between the two candidates who had received the most votes in the initial election. On Election Day, none of the candidates for either seat got a majority of the vote. Hence, runoff elections for both seats took place on January 5, 2021. These runoff elections drew nationwide attention because their outcome would determine which party would control the Senate. Democratic victories in both elections would give that party 50 seats. However, the vice president is constitutionally empowered to break ties in votes taking place in the Senate. Because newly elected Vice President Kamala Harris is a Democrat, by winning both Georgia seats the Democrats would gain majority control of the Senate.

Prior to having been elected to the Senate in 2014, Perdue had pursued a career in business, including stints as the chief executive officer of Reebok,

PillowTex, and Dollar General. However, he had never run for public office prior to his run for the Senate. He ran unopposed in the Republican primary for a second term in 2020. Perdue's major challenger was Democrat Jon Ossoff, who had been a Congressional staff member specializing in defense policy and foreign affairs and later became chief executive officer of Insight TWI, a television production company that produces documentaries about political corruption in various parts of the world. Ossoff, who is Jewish and by winning the election became the first Jewish Senator from Georgia, had never held public office, although he had run unsuccessfully for the U.S. House of Representatives in 2016. In the primary election in June, he won 52.8% of the vote in defeating six opponents. In the general election in November, Perdue received 49.7% of the vote while Ossoff won 48% and Libertarian Shane Hazel won 2.3%. Thus, Perdue fell short of the majority needed to win the seat without a runoff.

Loeffler had been the chief executive officer of Bakkt, a financial service provider firm prior to her appointment to Isakson's seat by Kemp. (Her husband, Jeffrey Sprecher, is head of the New York Stock Exchange. Hence, both she and Perdue are quite wealthy, in contrast to Ossoff and Warnock.) The now-incumbent Loeffler competed in the special election to serve the final two years of Isakson's term. This special election was held on the same day as the election for Perdue's seat and as the Presidential election. In contrast to regular Senate elections, party primaries for seats in special elections in Georgia do not take place. Rather, all candidates for the seat, regardless of party affiliation, run in a "jungle" primary with the two candidates receiving the most votes, regardless of party, advancing to a runoff if no candidate gets a majority of the vote.

In addition to Loeffler, five Republicans, eight Democrats, and six independent or third-party candidates competed in the primary. Loeffler's major Republican opponent was U.S. Representative Doug Collins. Collins's candidacy was backed by Trump and some other prominent conservative Republicans. Meanwhile, some Democrats became concerned that their votes would be split among their party's contenders to the extent that Loeffler and Collins would advance to the runoff. Accordingly, party leaders coalesced around the candidacy of Warnock, the senior pastor of Ebenezer Baptist Church in Atlanta. (Before his assassination in 1968, Martin Luther King, Jr. had also served in that position.) Warnock, who is Black, had become active in various political efforts in Georgia, including efforts to expand Medicaid and to promote voter registration. His candidacy was supported by former Presidents Carter and Barack Obama and by Biden, Harris, and other prominent public figures within Georgia and across the country.

In the November election, Warnock finished in first place with 32.9% of the vote. Before the election, public opinion polls had showed Loeffler and Collins running about even among Republican voters. However, Loeffler won 25.9% of the vote with Collins in third place with 19.9%, so Warnock and Loeffler advanced to the runoff. In combination, the eight Democratic candidates won 48.4% of the vote while the six Republicans combined for 49.4%.

Thus, the division between Democrats and Republicans in the special election paralleled the results of the election for Perdue's seat held on the same day. Because none of the candidates in either election won majorities in the November elections, runoff elections for both seats were scheduled for January 5, 2021.

The Runoff Senatorial Elections

With the two Senate seats in Georgia still undecided, leaders of both parties were well aware that if the Democrats won both seats then their party would control the Senate, whereas if the Republicans won either or both runoff elections then the GOP would retain control of the Senate. Hence, the two Georgia runoffs drew nationwide attention. Record amounts of money were spent in these elections. Nearly $500 million was spent on the race between Perdue and Ossoff, including both the November election and the runoff, making this the most expensive Senate campaign in U.S. history (Evers-Hillstrom 2021). More than $360 million was spent on the special election campaign, including the November regular election and the January 5 runoff.

After the November special election but before the runoff, Collins endorsed Loeffler. Trump also endorsed her and campaigned in person for both her and Perdue in December, although he was criticized for having focused more on his own narrow loss in the state than on the two Senate candidates in his speech (Suggs 2020). The other Democratic and Republican candidates who did not qualify for the special election runoff endorsed Warnock and Loeffler, respectively.

Given that both regular elections had been contested so closely, not surprisingly public opinion polls taken in December and early January predicted that both races would be very close. As might have been expected, nearly all Democrats supported Ossoff and Warnock while nearly all Republicans supported Perdue and Loeffler. For example, according to one poll, 97% of Democrats supported Ossoff and 88% of Republicans supported Perdue (University of Nevada-Las Vegas Lee Business School 2021). However, the Republicans came into the runoffs with two potential advantages. In both of the November regular elections, the Republican candidates outpolled their Democratic opponents. With his 49.7% of the November vote, Perdue had fallen just short of the majority that he would have needed to avoid a runoff. Supporters of the Libertarian candidate, Shane Hazel, whose presence in the race probably denied Perdue a majority of the vote in November, could have been expected reasonably to prefer the more conservative Perdue to the more liberal Ossoff.

The Republicans stood to benefit from lower turnout than in the presidential election in November. In general, turnouts in elections that do not coincide with presidential elections tend to be lower than in presidential elections. Turnouts tend to be particularly low among younger, urban, and minority residents, who tend to support Democrats, relative to older persons and residents of rural areas who are more likely to be Republicans and are generally more likely to vote. In addition, the latter parts of the runoff campaigns took place during the holiday season, when Georgia residents were likely to be less focused on politics. On the basis of these factors, some Republican leaders in Georgia and across the country expressed cautious optimism that Perdue and Loeffler would prevail in their respective runoffs.

All four candidates campaigned vigorously in the days leading up to the runoffs, and prominent supporters of each visited Georgia to promote their candidacies. The campaigns intensified during the second half of December, after early voting began three weeks before January 5. Trump visited Dalton on January 4 to campaign for Perdue and Loeffler. However, in a speech lasting more than an hour, Trump again focused on his own narrow loss of Georgia's electoral votes in November, including pointed criticism of Kemp and Raffensberger, both Republicans, for having refused to support overturning the results (Hallerman and Mitchell 2021). Outgoing Vice President Mike Pence and Trump's daughter Ivanka and son Donald, Jr. also campaigned for the two Republicans. On the Democratic side, President-elect Biden appeared at a rally in support of Ossoff and Warnock on December 15, the day that early voting began (Bluestein 2020). Vice President-elect Harris also visited Georgia to campaign in person for the two Democratic candidates while former President Obama, House Speaker Nancy Pelosi, and prominent Senator Elizabeth Warren headlined virtual rallies.

Republican optimism prior to the runoff elections proved to be unfounded as Ossoff and Warnock defeated their Republican opponents. Ossoff defeated Perdue by a margin of 54,946 votes, winning 50.6% of the vote. Meanwhile, Warnock defeated Loeffler by a margin of 93,272 votes, winning 51.0% of the vote. Perhaps reflecting the national importance of these elections and the record amount of money spent on the campaigns, turnout was nearly as high in the runoffs as it had been in the November elections. Turnout in the Ossoff–Perdue runoff was 90.6% of turnout in that race in November, while turnout in the Warnock–Loeffler runoff was 91.3% of turnout in that race in November. Some observers commented that the election of Georgia's first Black and Jewish Senators was a surprise, but party identification, the national importance of the outcomes, and stark differences between Democrats and Republicans on policy issues seemed to have overridden any such concerns (Weissman 2021).

The exit poll for the runoff between Ossoff and Perdue (CNN Politics 2020) showed sharp divisions between supporters of the two candidates. Not surprisingly, almost all Democrats voted for Ossoff and almost all Republicans supported Perdue. As well, Ossoff supporters were nearly unanimous in preferring Democratic control of the Senate nationwide and almost all reported having voted for Joe Biden in the 2020 presidential election. More than 60% of voters under 45 years of age supported Ossoff, while older voters preferred Perdue. As with many recent elections, a gender gap was evident in that a majority of women voted for Ossoff whereas a majority of men supported Perdue.

The Geography of the Senate Runoff Elections

Where did Ossoff and Warnock find the votes needed to win their respective runoff elections? The results of the two runoff elections were almost identical. Ossoff and Warnock each won 30 of Georgia's 159 counties, and each of the 30 counties that gave a majority of its votes to Ossoff also gave a majority of its votes to Warnock (Figures 5.2 and 5.3).

In recent presidential elections throughout the United States, a split between the Democrats in metropolitan areas and Republicans outside metropolitan areas has intensified (Niskanen Center 2019). In 2020, according to the national exit poll 60% of urban residents supported Biden whereas only 42% of rural residents supported him – a gap of 18%. Suburban voters were almost evenly divided, with 50% of voters supporting Biden and 48% supporting Trump. By contrast, in 2004 the urban–rural gap was only 12%, with 57% of rural residents supporting Republican George W. Bush whereas 45% of rural residents supporting him. The urban–rural split was even greater in the 2020 presidential election in Georgia, in which 67% of urban residents supported Biden whereas 69% of rural residents supported Trump. The percentages in the runoff between Ossoff and Perdue in January were identical.

Nine of the 28 counties carried by both Ossoff and Warnock are located in the twelve-county Atlanta metropolitan area. Ossoff and Warnock carried nine of these counties, losing in only the three smallest and most outlying counties in the region. The remaining 21 counties carried by Ossoff and Warnock include Richmond County (Augusta), Muscogee County (Columbus), Bibb County (Macon), Chatham County (Savannah), and Clarke County

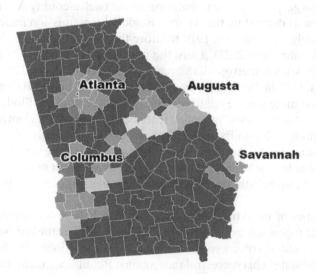

Figure 5.2 County-level percentage of vote for Ossoff and Perdue in runoff election of January 5, 2021.

Source: https://www.cnn.com/election/2020/results/state/georgia/senate-runoff

Figure 5.3 County-level percentage of vote for Warnock and Loeffler in runoff
   election of January 5, 2021.

Source: https://www.cnn.com/election/2020/results/state/georgia/senate-runoff

(Athens and the University of Georgia). Augusta, Columbus, Macon,
Savannah, and Athens are the five largest cities in Georgia that are located
outside the Atlanta metropolitan area.

The results of the Senate elections, as well as the result of the presidential
election, illustrate the degree to which the Atlanta area is increasingly dominant in Georgia politics. The population of the twelve-county Atlanta metropolitan area as defined by the Atlanta Regional Commission increased from
approximately 1.5 million in 1970 to more than 4.7 million in 2020 (Atlanta
Regional Commission 2021). Thus, the percentage of Georgia residents who
lived in the Atlanta metropolitan area increased from about 32% in 1970 to
more than 43% in 2020. Moreover, the electorate in the Atlanta area is
younger and more diverse relative to the rest of the state. As Slodysko (2021)
commented, Georgia has "morphed into a battleground – a change driven in
part by demographic shifts, particularly in the economically vibrant area of
metropolitan Atlanta. As older, white, Republican-leaning voters die, they've
been replaced by a younger and more racially diverse cast of people, many of
whom moved to the Atlanta area from other states – and carried their politics
with them."

The politics of the Atlanta metropolitan area have also changed in recent
years as the region has expanded. The 1992 election was the last one in which
Georgia's electoral votes went to the Democratic nominee with Bill Clinton
carrying the state in his successful race against Republican incumbent George
H.W. Bush. Clinton won Georgia by 13,714 votes – a slightly larger margin
than Biden's win over Trump in 2020. However, Clinton carried only three
of the twelve counties that now comprise the Atlanta metropolitan region.

He won Fulton County, which contains the city of Atlanta and has a large Black population, with 57% of the vote. He also won adjacent DeKalb and Clayton Counties, with 58% in DeKalb County and 45% in Clayton County, where he outpolled Bush because nearly 14% of the vote in that county went to independent candidate Ross Perot. Statewide, Clinton carried 92 of Georgia's 159 counties. By contrast, in the January runoff Ossoff received 72% of the vote in Fulton County, 84% of the vote in DeKalb County, and 88% of the vote in Clayton County. The percentages of the vote in these and other counties in the Ossoff–Perdue race across the state were almost identical in the Warnock–Loeffler runoff election.

The contrast in the suburbs was even more striking. Historically, Gwinnett County northeast of Atlanta was a conservative stronghold. In the 1992 presidential election, Bush won 54% of the vote as compared to 30% for Clinton and 16% for Perot. In the 2021 runoff, however, Ossoff got 60% of Gwinnett County's vote to 40% for Perdue. Moreover, nearly 370,000 votes were cast in the runoff in Gwinnett County in 2021, whereas approximately 150,000 were cast in the 1992 presidential election. Not only did Gwinnett County shift dramatically from the Republicans to the Democrats, but its population increased by 170% during the three decades between 1990 and 2020. This population increase is reflected in the increase in voter turnout. Demographic and electoral changes in Gwinnett County were consistent with such changes throughout the Atlanta metropolitan area. The population of each of these twelve counties grew by more than 40% between 1990 and 2020. Gwinnett County was one of four counties in the region whose population more than doubled during this three-decade period.

Rapid population growth and shifts toward the Democrats in the Atlanta metropolitan region is contrasted with changes in many counties in other parts of the state. For example, Perdue's share of the vote exceeded 77% in each county in a region of southeastern Georgia, including Appling, Bacon, Brantley, Jeff Davis, Pierce, and Wayne Counties. However, of these six counties only Brantley County experienced a population increase of more than 50% between 1990 and 2020. Moreover, the populations of these counties are small; for example, Appling County's population was 18,106 and Bacon County's population was 11,228 in 2020. The combined total population of these six counties in 2020 was 113,720 as compared to Gwinnett County's 2020 population of 954,076. This example illustrates the degree to which Republican candidates for statewide and national offices remain highly successful in rural counties, but that these counties have small populations with modest growth rates relative to those in the Atlanta metropolitan area. In 1990, these six counties had a combined population of 84,103 as compared to 352,910 in Gwinnett County.

The major exceptions to the general split between urban support for Democrats and rural support for Republicans occurred in rural counties with large Black populations. For example, Macon County in west-central Georgia (not to be confused with the city of Macon, which is located in Bibb County) gave 60.5% of its votes to Warnock and 59.4% of its votes to Ossoff. Nearly

60% of Macon County's residents are Black. Dougherty County in southwestern Georgia gave 68% of its votes to both Democrats, and it has a population that is 71% Black. On the other hand, none of the six contiguous counties in southeastern Georgia that gave more than three-quarters of their votes to both Perdue and Loeffler have populations that are more than 20% Black. More generally, in rural Georgia there was a very close relationship between Black population percentages and levels of support for both Ossoff and Warnock.

In majority-Black counties and throughout the state, support for Warnock among Black voters was particularly noteworthy in the runoff election. Black voters' enthusiasm for Warnock, who, by winning his runoff election, became the first Black Senator elected from Georgia, has been cited as a reason why Warnock won his seat by a larger margin than did Ossoff (Cohn 2021). Warnock defeated Loeffler by a margin of 93,272 votes whereas Ossoff defeated Perdue by a margin of 54,944 votes. However, Ossoff emphasized outreach to Black voters in his campaign. He also called attention to his having been endorsed by John Lewis, the iconic civil rights leader and long-time member of the House of Representatives from Atlanta, before Lewis died in July 2020. Arkin and Desiderio (2021) pointed out that "In the first week of this past December [2020], Ossoff drove to Selma, Alabama, to film a TV ad on the historic Edmund Pettus Bridge, invoking Lewis' history as a civil rights leader and calling for a new Civil Rights Act." Black voters have been recognized as crucial to the success of both Democrats in the runoff elections (Arkin and Desiderio 2021).

The Republican Dilemma

By winning their Senate seats in the January 2021 runoff elections, Jon Ossoff and Raphael Warnock flipped the two Georgia seats from the Republicans to the Democrats and in doing so ensured that the Democratic Party would control the Senate in light of incoming Vice President Harris' authority to cast tie-breaking votes. Of course, Warnock's seat will be up for election in November 2022. Frequently, midterm elections represent measures of public opinion about the current presidential election. Hence, if the Biden–Harris administration and/or the performance of the current Congress are unpopular in Georgia by the summer and fall of 2022, this might bode poorly for Warnock's chances of being elected to a full term. It is also possible that Loeffler, Perdue, or another prominent Republican with statewide name recognition might choose to challenge Warnock. As well, in early 2021 after the runoff elections, the Republican-controlled legislature of Georgia enacted laws that would restrict access to voting in future elections, including more rigid voter identification requirements (Corasaniti 2021). It has been predicted widely that such laws, if upheld by the judiciary, will have a disproportionate impact on poor and minority voters who are much more likely to vote for Democratic candidates for public office.

But in the longer run, the Democrats' success in the two Georgia Senate elections may foreshadow a dilemma facing the contemporary Republican

Party in Georgia and across the United States. As we have seen, Ossoff and Warnock won their elections on the strength of support among younger and more diverse voters living in the Atlanta metropolitan area whereas the two losing Republican candidates, Perdue and Loeffler, did best among older white voters living in rural areas. Demographic changes, along with in-migration to the Atlanta area from other states, suggest that the continuation of present trends will bode well for the Democrats going forward. Similar trends have been observed in Texas, North Carolina, and other Southern states with large, growing metropolitan areas. Some observers predict that these states may also move into the Democratic column in the foreseeable future.

Perhaps in response to the victories of Biden, Ossoff, and Warnock, early in 2021 the state's Republican-dominated legislature enacted highly restrictive voting laws. These laws, if upheld by the judiciary, are expected to have disproportionate effects on Black voters and would therefore benefit Republicans in upcoming elections (Fausset et al. 2021). Above and beyond efforts to restrict voting, what can the Republicans do in response to these continuing demographic and political changes? Several months after he left office, Donald Trump remains very popular among the Republican rank and file. On the other hand, Trump's unpopularity among moderate suburban voters contributed significantly to his loss in Georgia and other closely contested states – losses that cost him a second term despite his continued protestations that the election was stolen from him. To what extent might Republicans in Georgia reach out to these more moderate voters as opposed to strong adherence to Trump and his supporters going forward, even at the risk of losing challenges to Warnock and Ossoff as they run for re-election?

References

Arkin, J. and A. Desiderio. 2021. How Warnock and Ossoff painted Georgia blue and flipped the Senate. *Politico* (Jan. 7). https://www.politico.com/news/2021/01/07/warnock-ossoff-flipped-senate-georgia-456310.

Atlanta Regional Commission. 2021. About the Atlanta region. https://atlantaregional.org/atlanta-region/about-the-atlanta-region.

Bluestein, G. 2020. Biden presses Democrats to ensure runoffs aren't "even close". *Atlanta Journal-Constitution* (Dec. 15). https://www.ajc.com/politics/politics-blog/live-updates-biden-is-set-to-rally-for-runoff-candidates-in-georgia/QBITV2R4URDTJAOLLOI3OP2F2Q/.

Buchanan, S. 2005. County unit system. *New Georgia Encyclopedia.* https://www.georgiaencyclopedia.org/articles/counties-cities-neighborhoods/county-unit-system.

CNN Politics. 2020. Exit polls. https://www.cnn.com/election/2020/exit-polls/senate-runoff/georgia.

Cohn, N. 2021. Why Warnock and Ossoff won in Georgia. *New York Times* (Jan. 7). https://www.nytimes.com/2021/01/07/upshot/warnock-ossoff-georgia-victories.html.

Cohn, N., M. Conlen, and C. Smart. 2020. Detailed turnout data shows how Georgia turned blue. *New York Times* (Nov. 17). https://www.nytimes.com/interactive/2020/11/17/upshot/georgia-precinct-shift-suburbs.html.

Corasaniti, N. 2021. Georgia G.O.P. passes major law to limit voting amid nationwide push. *New York Times* (April 3). https://www.nytimes.com/2021/03/25/us/politics/georgia-voting-law-republicans.html.

Evers-Hillstrom, K. 2021. Georgia Senate races shatter spending records. opensecrets. org (Jan. 4). https://www.opensecrets.org/news/2021/01/georgia-senate-races-shatter-records/.

Fausset, R., N. Corasaniti, and M. Leibovich. 2021. Why Georgia G.O.P.'s voting rollbacks will hit Black people hard. *New York Times* (March 25). https://www.nytimes.com/2021/03/25/us/politics/georgia-black-voters.html.

Fowler, S. 2020. Judge denies last-minute request to block Georgia election certification. *GPB News* (Nov. 19). https://www.gpb.org/news/2020/11/19/judge-denies-last-minute-request-block-georgia-election-certification.

Gardner, A. 2021. "I just want to find 11,780 votes": In extraordinary hour-long call, Trump pressures Georgia secretary of state to recalculate the vote in his favor. *Washington Post* (Jan. 3). https://www.washingtonpost.com/politics/trump-raffensperger-call-georgia-vote/2021/01/03/d45acb92-4dc4-11eb-bda4-615aaefd0555_story.html.

Hallerman, T. and T. Mitchell. 2021. Trump campaigns for Loeffler and Perdue in high stakes election eve rally. *Atlanta Journal-Constitution* (Jan. 4). https://www.ajc.com/politics/live-trump-campaigns-for-loeffler-and-perdue-in-high-stakes-election-eve-rally/R7E5SMX2RJALPNS7JFZ65W6WTY/.

Niskanen Center. 2019. Explaining the urban-rural political divide. (July 17). https://www.niskanencenter.org/explaining-the-urban-rural-political-divide/.

Slodysko, B. 2021. How democrats won Georgia's 2 Senate runoffs. *Associated Press* (Jan. 6). https://apnews.com/article/associated-press-georgia-election-result-60954fd7d3d3b6b49a8884c0c026247d.

Suggs, D. 2020. Trump campaigned for Loeffler and Perdue – and himself – on Saturday night in Georgia. *MarketWatch* (Dec. 6). https://www.marketwatch.com/story/trump-campaigned-for-loeffler-and-perdue-and-himself-on-saturday-night-in-georgia-11607264400.

University of Nevada-Las Vegas Lee Business School. 2021. Georgia State Primary Poll. Jan. 4. https://busr.ag/georgia-senate-poll.

Weissman, S. 2021. Scholars reflect on the symbolism of a Black-Jewish Senate election win in Georgia. *Diverse Issues in Higher Education* (Jan. 7). https://www.diverseeducation.com/home/article/15108444/scholars-reflect-on-the-symbolism-of-a-black-jewish-senate-runoff-election-win-in-georgia.

6 Covid-19, Race, and the 2020 Election in Wisconsin

Ryan Weichelt

Since it was thrust into the national spotlight in 2011, Wisconsin has embodied the partisan political climate seen across much of the United States. The success of Tea Party candidates in Wisconsin after the 2010 midterms helped lay the foundation for Donald Trump's victory in the years that followed. Starting in 2011, Republicans systematically altered the political landscape of Wisconsin through legislation and blatant gerrymandering, ensuring majorities in the State Senate and Assembly as well as maintaining a majority of Wisconsin House seats. The newly elected Republican Governor Scott Walker initiated a conservative agenda that permeated Democratic voting districts across rural Wisconsin. Starting with his key Act 10 legislation that curtailed the collective bargaining rights of public employees, Walker would continue to push a discourse in the years ahead that consistently highlighted an urban versus rural divide. This discourse was further pushed at the federal level by characters like Senator Ron Johnson, who won re-election in 2016, and Representative Sean Duffy (House District 7), both of whom eventually became some of Trump's loudest cheerleaders.

The 2016 presidential election sent the clear message to Democrats nationwide that their party had lost the ability to connect to rural, blue-collar white voters, particularly in the Midwestern states of Wisconsin and Michigan. The disconnect between the Democratic Party and persons outside of Wisconsin's urban and suburban regions only widened between 2010 and 2016. Rural Wisconsin, in particular, is hampered by a crumbling infrastructure, consistent poverty, struggling schools, and unemployment. It could be argued these issues would resonate stronger with Democratic ideologies versus Republican; however, at the state level, Wisconsin Democrats had done little to connect with rural voters in the last decade. After their sweeping victories of the Governorship and Legislature in 2010, Wisconsin Republicans implemented legislation to increase hunts on gray wolves, relax deer hunting regulations, and introduce an abstinence-only sex education curriculum for public schools within the state. The agenda was designed to both appeal to rural voters and distract them from the poor economic realities of the post-2008 recession.

Perhaps no other state embodies these electoral divides and has influenced recent national politics more than Wisconsin. Wisconsin had long been a Democratic stronghold for presidential elections, having not elected a

DOI: 10.4324/9781003260837-6

Republican candidate for president since 1984. Combined with the other Democratic strongholds of Michigan and Pennsylvania, pundits aptly named these three states the "Blue Wall." Collectively, this wall has held for Democrats in every presidential election since 1992 and few believed it would fall in 2016. In the weeks leading up to the election, Hillary Clinton held a comfortable lead in the polls throughout the Blue Wall states, but polls in Wisconsin indicated Clinton had the largest lead of the three. Riding success in the polls, Clinton made the decision to not visit Wisconsin during the general election. As the Blue Wall crumbled, many pointed to Wisconsin as the biggest surprise in 2016.

Having elected some colorful individuals in the past (Robert Lafollette and Eugene McCarthy to name a few), Trump can now be added to this list of characters. Though after the surprise of 2016 faded and analysis was conducted, Donald Trump's victory in the state, propelled by rural voters, should have been no surprise. The divide between urban and rural areas persisted into the 2018 and 2020 election cycles, although the state swung to the left in the presidential and gubernatorial races. Therefore, analyzing Wisconsin's recent election outcomes and the events leading up to the November 2020 election in the Badger State provides key components in understanding Joe Biden's eventual victory, not only in Wisconsin but nationwide.

Strange Bedfellows: Donald Trump and Scott Walker

Conflicts between urban and rural populations are not new in American politics. The Populist Movement of the late 1800s pitted agrarian interests of the Great Plains against the growing wealth of the northern and eastern industrializing states. Bonikowski (2016) suggests populism should be seen as a thinly centered ideology vilifying elites and glorifying people serving special interests. Therefore, populist leaders can capitalize on public dissatisfaction, fear, and resentment to serve their own political agenda. In present-day Wisconsin, Cramer (2016) exposed long-held rural resentment toward the large urban areas of Milwaukee and Madison in her book *The Politics of Resentment*. Cramer identified three elements in her analysis: a belief that rural areas are ignored by decision-makers, rural areas do not get their fair share of resources, and rural populations have a distinct culture that is both misunderstood and ridiculed by urban populations. Scott Walker and, later, Donald Trump represent, according to Shafer (2015), the "standard American demagogue" who relies on anger and resentment to attract supporters. The success of this rhetoric allowed Republicans to gain support across much of the state over the decade, culminating in Trump's election victory in 2016.

In 2015, Scott Walker was considered an early favorite as the Republican nominee for president. His bold legislative initiatives had brought him national attention. Though Trump was attracting interest, he was seen by many as a sideshow at the time. Trump's bombastic style contrasted Walker's carefully scripted and "good guy" demeanor. After a report surfaced of a Walker fundraiser calling Trump a "DumbDumb," Trump lashed out (Campbell 2015).

He relentlessly attacked Walker in the first two debates, suggesting Wisconsin was a disaster under his leadership. Unable to mount a defense against Trump or to distinguish himself from the rest of the field, Scott Walker was the first major candidate to drop out of the race (Cillizza 2019). In the months ahead, Walker and Trump continued to feud. This battle reached a crescendo in the Wisconsin primary, where Walker actively campaigned for Ted Cruz. Trump tweeted "After the way I beat Gov. Scott Walker (and Jeb, Rand, Marco and all others) in the Presidential Primaries, no way he would ever endorse me" (Gass and Glueck 2016). Although Cruz won the Wisconsin primary, Trump would eventually win the Republican Party's nomination. In the lead-up to the November 2016 election and in the aftermath of Trump's surprising victory, Walker and many other establishment Republicans had little recourse but to accept Donald Trump as the new leader of the GOP.

By 2017 both Trump and Walker were lagging in the polls. Walker's approval rating dipped to 48% and Trump's hovered around 42% (Paquette and Frankel 2019). Though once adversaries, Walker reluctantly and necessarily courted Trump. Sensing an opportunity, Walker sat down with President Trump and Terry Gou, CEO of Taiwan-based Foxconn, in the Oval Office in late April. By July 26, 2017, in the East Room of the White House, Donald Trump and Scott Walker jointly announced an unprecedented deal with Foxconn to build a massive LCD display manufacturing plant in southern Wisconsin. The deal provided Foxconn $4.1 billion in taxpayer-based subsidies; in return, the company promised it would lead to the creation of 13,000 jobs and an overall investment of $10 billion into the project (Murphy 2018). Wisconsin Democrats were livid. Foxconn was known for backing out of a similar deal in Pennsylvania in 2013 and the company also promised large investments in Indonesia, Vietnam, Brazil, and India but never delivered. Critics further argued the large investment in state dollars would mean Wisconsin taxpayers would be paying back the loan given to Foxconn well into the 2030s (Linnane 2017). Despite this, both leaders vied for credit for landing the deal. Though Walker and Wisconsin Republicans initiated the deal, Trump seized the opportunity to take credit for to contract that would highlight his America First campaign promise. In a 2018 campaign rally for Walker in Mosinee, Wisconsin, Trump stated "I got him [Walker] to set up with the incredible company called Foxconn ... I handed it over to Scott" (Levin and Busis 2018). Though initially reluctant to attach themselves to Trump, Republicans knew their future success was connected to the president. After the Foxconn deal, Walker and Republicans seized every opportunity to be seen with Trump before the June 2018 groundbreaking ceremony in Pleasant Prairie, WI. The Foxconn deal would both cement Walker's legacy in Wisconsin as well as forever connect Walker to Trump in a deal the Taiwanese company would never deliver.

As the 2018 midterm elections approached, Democrats and political pundits alike predicted a massive Democratic "blue wave" would result in victories across the United States. Up to the midterms, Donald Trump had secured only one major legislative victory (tax cuts) and was under strict scrutiny by the press at every moment. Further hampering Trump was the ongoing Mueller investigation into Russian interference in the 2016 election. In Wisconsin, Walker was facing his fourth election (2010, 2012 recall election, and 2014) in eight years and popular progressive Democratic Senator Tammy Baldwin was also up for re-election. In both races, the incumbents faced candidates currently holding elected office in the state. Baldwin faced state senator Leah Vukmir and Walker faced the quirky state Superintendent of Schools, Tony Evers. Walker himself warned Republicans after Wisconsin's April 2018 elections in a tweet: "Tonight's results show we are risk of a Blue Wave in WI" (Lafond 2018). For Democrats, the campaign was easy: connect both candidates to Donald Trump and Foxconn.

The 2018 Wisconsin Gubernatorial Election

For many first-term presidents, midterm elections generally have served as a reaction to the party of the president. Ronald Reagan saw Republicans lose 26 seats in the House and both Bill Clinton and Barack Obama saw Democrats losing majority control during their first midterm elections. The 2018 midterms were poised to be no different for Donald Trump than his predecessors. Republican candidates were facing not only a highly energized Democratic electorate, they were seemingly also running against Donald Trump himself. The 2016 election saw Donald Trump energize large swaths of voters across rural America who normally had not voted in previous elections. The consensus after the election is that pollsters did not apply enough statistical weight to poll respondents without college degrees and that they overestimated broad support for Clinton (White 2020). For 2018 Republican candidates, it was unknown if the energized Trump base that emerged in 2016 would turn out in 2018 without Trump's name on the ballot. Due to Wisconsin's position as a state that flipped in 2016 for Trump, the 2018 midterm elections in Wisconsin provide an excellent electoral landscape to measure Trump's reach.

As Walker reluctantly allied himself to Trump, Wisconsin Democrats saw an opportunity to dethrone their greatest antagonist. Beyond the growing national contempt for Donald Trump, Wisconsin Democrats had been gaining momentum in 2018. In a non-partisan election, liberal Wisconsin Circuit Court judge Rebecca Dallet easily won a 10-year term to the State Supreme Court in April. In the lead-up to the August 2018 Democratic primary, it was anybody's guess who would be selected as the candidate to face Walker. Moderate Democrats floated long-serving Congressman Ron Kind as the best candidate, while Progressives urged Madison Congressional Representative Marc Pocan to run. By July, over 14 people had declared their intentions to run, including a 2012 candidate for governor, Kathleen Vinehout, and popular Madison Mayor Paul Soglin. Though a quirky candidate would emerge from the crowded field, Tony Evers.

Tony Evers grew up in the small city of Plymouth, Wisconsin in central Sheboygan County. After graduating from the University of Wisconsin-Madison, Evers began his career as a teacher in Tomah, Wisconsin, and eventually settled into administration at a variety of school districts through-out southern Wisconsin. Though many described Tony Evers as quirky and bland, Evers ran for political office in 1993 and 2001 for State Superintendent of Public Instruction, losing both (Davey 2018). However, after the election in 2001, Evers was appointed as the Deputy of Public Instruction by Elizabeth Burmaster. In 2009, after Burmaster took a job elsewhere, Evers won the statewide nonpartisan election for State Superintendent. The eventual victory by Walker in 2010 and the debate centered on Walker's signature legislation, Act 10, forever linked the two men as adversaries. As 70,000 protestors settled on the steps of the state capitol in Madison protesting Walker's attack on public unions and teachers, Evers gained statewide recognition. Though Evers could do little to stop Act 10, his position as lead educator engrained his image to citizens across the state. Evers easily won re-election in 2013 and again in 2017, with over 70 percent of the total vote. As the 2018 Democratic gubernatorial primary came to an end, and despite beliefs Evers was not "inspiring" enough to defeat Walker, the quirky teacher won a surprising 40 percent of the total vote among seven candidates.

Throughout the campaign, Evers was quick to attack Walker on a variety of topics, from roads to the environment to healthcare, all of which Walker could brush off, but education, Foxconn, and Trump were topics Walker could not avoid. The first line of Evers's first campaign ad starts "The Donald Trump-Scott Walker $3 billion deal for Foxconn might sound good, until you look at the fine print." Evers later said in the ad "Sounds good for Foxconn, but what's in it for the rest of us? Just think if we invested that money in our schools instead" (Sommerhauser 2017). Throughout the campaign the Wisconsin governor did not have a strong response and did little to invite new voters to his camp. In the final weeks of the campaign, Walker's message was perhaps overly simple, asking voters to "Let's finish what we started" (Marley and Crowe 2018).

As the polls closed at 8 p.m. on November 6, 2018, the governor's race was deemed "too close to call." Early results showed Walker in the lead, but by around 10 p.m., as Dane County's vote share was reported, the election result was nearly even. By midnight, a winner was still not determined, but when Brown County finally submitted its tally, Walker held a slight 10,000 vote lead. It seemed Walker had survived yet again, but by 12:45 a.m., Milwaukee County Clerk George Christenson said they received around 46,000 early and absentee ballots from the City of Milwaukee that were processed after all the same day vote was counted. Evers received 84 percent of these votes, giving him around a 31,000 vote lead (Bice and Spicuzza 2018). In a somewhat surprising result, Walker had been defeated.

By the next day, though some in the Walker camp suggested legal action may be taken, Scott Walker offered his concession to Tony Evers. Statewide, Democrats won every election. Figure 6.1 shows the results of the election at both the county and the voting district level. Evers' victory was due in large part to three main factors: large support in Dane County, low turnout in rural areas, and lower level of support for Walker in the traditional Republican

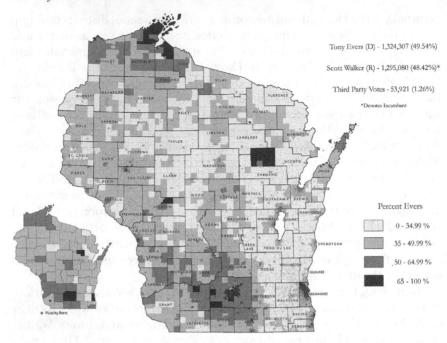

Tony Evers (D) - 1,324,307 (49.54%)

Scott Walker (R) - 1,295,080 (48.42%)*

Third Party Votes - 53,921 (1.26%)

*Denotes Incumbant

Percent Evers

0 - 34.99 %

35 - 49.99 %

50 - 64.99 %

65 - 100 %

Figure 6.1 Percent Tony Evers Vote for the 2018 Gubernatorial Election at the County and Voting District Level.

Sources: map by author; data from Wisconsin State Legislature and Wisconsin Election Commission.

stronghold of Waukesha County. In the three previous gubernatorial elections with Scott Walker on the ballot, Dane County offered between 81,000 and 102,00 extra votes to the Democratic candidate. In 2018, Dane Democrats provided Tony Evers a 150,808 margin of victory over Scott Walker. In Waukesha County, a typical Republican stronghold, Walker's margin of victory was 74,572, the lowest of any of his previous elections (Kremer and Johnson 2018).

This election clearly demonstrated Trump voters were less interested in voting when he was not on the ballot. Figures 6.2 and 6.3 show the voting district results comparing the percent difference in votes for Democratic and Republican candidates between the 2018 and 2016 elections. As can be seen, support was depressed throughout the rural areas compared to 2016. Democratic voters in rural areas exceeded their levels compared to 2016 and were more excited in urban areas across the state. In the suburban voting districts, Donald Trump was not nearly as popular in 2016 compared to rural voters. Figures 6.2 and 6.3 show Walker overperformed Donald Trump in these areas, but the lower-than-normal support in Waukesha County and other suburban communities around Milwaukee, the Fox River Valley, and St. Croix County in Western Wisconsin contributed to Walker's only defeat as governor.

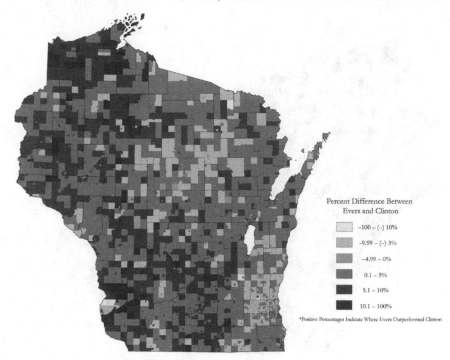

Figure 6.2 Percent Difference Between Tony Evers and Hillary Clinton at the Voting District Level.

Sources: Map by author; data from Wisconsin State Legislature and Wisconsin Election Commission

The 2020 Wisconsin Primary and Covid-19

Nationwide Democrats were reinvigorated by the 2018 midterms. Though many believed the "Blue Wave" would be bigger, Democrats did take back control of the House of Representatives and made modest gains in the U.S. Senate. With a newfound power in the House, Democrats, led by Nancy Pelosi, were finally able to go on the offensive. Trump continued to hurt himself through his bombastic comments and continued use of misinformation. His greenlighting of white supremacists in the aftermath of the Charlottesville protest and his aggressive stance against undocumented immigrants provided plenty of fodder for Democrats. Yet the Mueller investigation would weigh the heaviest for Trump. Democrats attacked Trump relentlessly and by the end of 2019, Donald Trump became only the third president of the United States to be impeached by the House of Representatives.

With a renewed sense of urgency to defeated Trump, 29 Democrats across the political spectrum vied to challenge Trump in 2020. Headliners included the 2016 candidate Vermont Senator Bernie Sanders, Massachusetts Senator Elizabeth Warren, Minnesota Senator Amy Klobuchar, California Senator

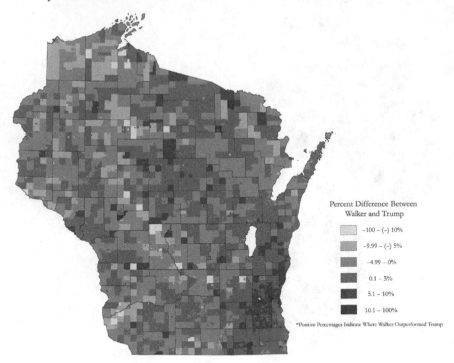

Figure 6.3 Percent Difference Between Scott Walker and Donald Trump at the Voting
District Level.

Sources: Map by author; data from Wisconsin State Legislature and Wisconsin Election
Commission

Kamala Harris, former South Bend, Indiana Mayor Pete Buttigieg, and former Vice President Joe Biden. Though Trump was the focus, the campaign
highlighted cracks in the Democratic Party between Progressive candidates
and traditional moderate Democrats.

As December gave way to January, all eyes were focused on the Iowa caucuses, but in the early weeks of the new year, stories of the highly contagious
virus Covid-19 – initially known as the coronavirus – in China started to
permeate the nightly news.

On January 21, 2020, the first case was identified in the United State in
Washington state from a man who had visited Wuhan a week earlier. Though
attention to the virus was largely muted by Trump's first impeachment and
the forthcoming presidential primary, Covid-19 continued to spread. On
January 30, WHO declared Covid-19 a "Public Health Emergency" and the
next day Trump would ban travel from China.

As more attention was being placed on the emergence of Covid-19 in the
United States, Pete Buttigieg surprised many by narrowly defeating Bernie
Sanders in the Iowa caucuses in early February. Seven days later, Buttigieg
surprised once again by nearly defeating Bernie Sanders in the New

Hampshire primary. Elizabeth Warren and Joe Biden vastly unperformed in both contests, but Biden was not deterred. The former vice president placed his early bets on a Southern strategy, relying on his reputation among African American voters in South Carolina on February 29 and the Super Tuesday primaries on March 3, when many Southern states held their primaries. Biden's strategy worked. The former vice president easily won the South Carolina primary and, as polls closed on March 3, Biden would win ten of the fifteen elections in the South. On March 10, Biden won five more states, placing him comfortably ahead of Bernie Sanders.

Yet, with Biden gaining clear momentum, the reality of Covid-19 quickly took hold in the United States. On March 11, the WHO declared Covid-19 a pandemic and on March 13 President Trump declared a national emergency. By April 1, nearly every governor across the county closed schools and initiated lockdown procedures, hoping to halt the spread. It was unknown how long lockdowns would continue and what impacts Covid-19 would have not only on the primary election, but on the everyday lives for Americans in the months ahead.

The 2020 Wisconsin Primary

Since Tony Evers's surprising victory in 2018, Republicans across the state were stunned and angry. Having gone unchecked for eight years, Wisconsin Republicans were now left with few options to pass their agenda. In a last-ditch attempt to yield some power, Republicans and outgoing governor Scott Walker passed legislation in mid-December 2018 stripping some of the governor's powers. The new laws would "curb the authority of Governor (Evers) in the rule-making process, limit early voting, allow legislators to intervene in some lawsuits, and limit the power of the Attorney General" (Smith and Davey 2018). This tit-for-tat continued into Evers's first year: Evers would veto, and both sides would squabble. By March, 2020, it was unknown if the pandemic would bring both sides together.

On March 13, Governor Tony Evers ordered all schools to be closed by March 18. On March 24 he issued a "Safer at Home" order effective for thirty days, asking residents to stay at home as much as possible to curb the spread of Covid-19. On the same day nine states set to hold primaries in March and April postponed their elections and/or decided to hold the elections entirely by mail. In Wisconsin, Evers's "Safer at Home" order put the April 7 primary in jeopardy. Initially, Evers and Republican leaders insisted an in-person election would take place, but in the coming days, as Covid infections increased across the state, Evers changed his position. The Democratic governor asked the GOP-led legislature to move the election entirely to voting by mail.

Evers's suggestion to move the election to all-mail voting would exacerbate growing Republican concerns about absentee ballots. As Covid continued to spread, it was obvious that mail-in voting would play a larger role than it ever had in any previous election. In Washington, DC, Democrats attempted to include universal vote by mail language in the first Covid stimulus package signed on March 27, but the proposal was later removed. In Wisconsin, Evers and Democrats attempted to use public safety to hinder the spread of Covid to strengthen their case for an all-mail primary. Wisconsin Republicans scoffed at the idea, suggesting it was logistically impossible to move the entire election to a mail-in affair, but said little about the safety of voting in person or offer any solutions. Trump himself would further suggest without evidence, on March 31 on *Fox and Friends*, that absentee voting increases voter fraud and that such a system would never allow a Republican to be elected again (Queally 2020). Former U.S. Attorney General Eric Holder prophetically stated that situation in Wisconsin regarding the primary election was "a microcosm. And presents questions that the nation will soon have to grapple with" (Corasaniti and Saul 2020).

It was clear that Covid was quickly transforming into a partisan issue. After Wisconsin Republicans refused Evers's request for a mail-in only primary, Evers requested the primary be postponed. Republicans refused and the issue went to federal court on April 2. In his decision, Judge William Conley rebuked both sides for turning the pandemic into a partisan fight. Conley stated "As much as the court would prefer that the Wisconsin Legislature and Governor consider the public health ahead of any political considerations, that does not appear in the cards." He also believed a federal judge should not interfere saying it is not, "appropriate for a federal district court to act as the state's chief health official by (deciding if an election should be held or not)" (quoted in Richmond 2020).

Conley's decision would uphold the in-person primary on April 7, but the judge also understood, due to the virus, use of absentee voting would exceed any previous levels increasing stress on local officials. Therefore, he ruled ballots could be accepted by election clerks up to April 13 by 4 pm; Republicans were infuriated. Evers attempted to hold a special session of the Legislature regarding the issue, but Republicans refused to talk and ended the session after only 17 seconds. Arguing that the Conley decisions effectively moved the election back one week, Republicans appealed Conley's ruling to the United States Supreme Court (Richmond 2020). On April 6, in a partisan 5–4 decision, the U.S. Supreme Court ruled the election had to be completed on April 7 at 8 p.m., but the Court did allow any ballots postmarked by April 7 to be accepted up to April 13. Immediately after the ruling, Evers issued an Executive Order postponing the election, but Republicans appealed the ruling to the Wisconsin State Supreme Court. In another partisan decision, in the evening of April 6, the conservative majority ruled Evers exceeded his authority, and the election would take place on April 7.

In the weeks leading up to Wisconsin's primary, it was becoming clear Joe Biden would be the Democratic nominee. Yet Wisconsin's voters would also be voting for a non-partisan ten-year seat on Wisconsin Supreme Court and a variety of local elections. With the election set for April 7, most Wisconsinites would choose the safety of the mail over in-person voting. In an unprecedented turn of events, Wisconsin voters would take to voting by mail with nearly 74 percent of total ballots cast as absentee. The 1.1 million votes cast absentee were a dramatic increase from the previous high of 140,000 absentee ballots cast during the 2016 Presidential Election (Root 2020).

The results of the primary would provide insight into previous elections and foreshadowing for the fall. Liberal State Supreme Court candidate, Jill Karofsky, easily defeated Walker-appointed Justice Daniel Kelly. This result (Figure 6.4) mirrored Tony Evers' 2018 surprise victory and showed continued gains for Democrats in suburban areas across the state. As the pandemic and absentee voting took a highly partisan tone, Democrats were more likely to embrace voting by mail compared to Republicans. Figure 6.5 shows the proportion of the total vote that was absentee and, when coupled with Figure 6.4, results showed that counties with the higher proportions of the total vote cast via the mail, generally correlated with higher support for Karofsky. Fortunately for many Wisconsin voters who decided to vote in person on April 7, few cases of Covid were attributed to the primary election. Though the fight regarding absentee ballots and Covid that started Wisconsin had only just begun.

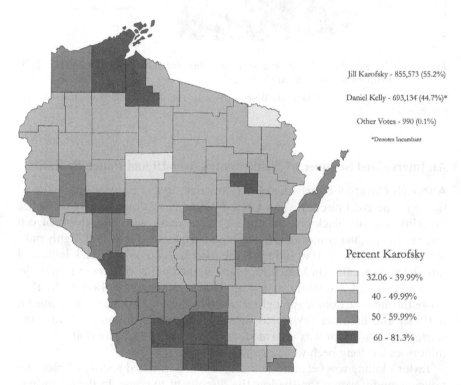

Jill Karofsky - 855,573 (55.2%)

Daniel Kelly - 693,134 (44.7%)*

Other Votes - 990 (0.1%)

*Denotes Incumbant

Percent Karofsky

32.06 - 39.99%

40 - 49.99%

50 - 59.99%

60 - 81.3%

Figure 6.4 Percent Vote for Jill Karofsky, 2020 Wisconsin State Supreme Court Election for Wisconsin Counties.

Sources: Map by author; data from Wisconsin State Legislature and Wisconsin Election Commission

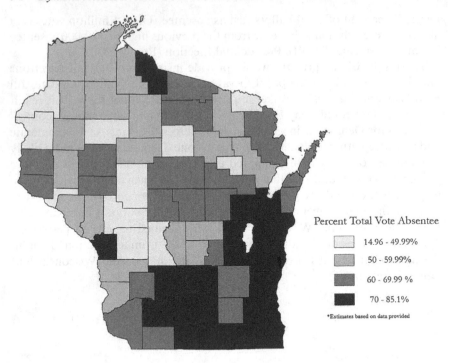

Figure 6.5 Percent of the Total Vote that was Absentee Ballots for the April 7, 2020 Primary for Wisconsin Counties.

Sources: Map by author; data from Wisconsin State Legislature and Wisconsin Election Commission

An Intertwined Summer of Discontent: Covid-19 and Police Brutality

Although Covid-19 clearly disrupted America, it was not the only issue to figure in the 2020 election. The United States has a long history of police brutality towards black citizens and in recent years, with the advent of cell phone cameras, many more examples have been made public. Yet highly publicized incidents in 2020 would push the issue beyond only a Black issue and into the mainstream. On March 13, plain-clothed police officers in Louisville, Kentucky served a warrant at the apartment of Breonna Taylor. As they broke through the door a gun fight ensued. In the end, a police officer was hit in the leg and Breonna Taylor was shot six times and pronounced dead at the scene. Though Taylor was unarmed, excessive use of force by police against minorities has long been well documented.

Taylor's killing was yet another dark stain in a troubled history of race and policing, and it would foreshadow the discontent to come. In the meantime, the pandemic and the stress of the lockdowns began to grow. Across the country, and especially in Wisconsin, lockdowns would not only become a partisan issue, but they also exacerbated already strained rural and urban divides. As Governor Evers's initial "Safer at Home" order was set to expire by April 24, he directed the Department of Health Services (DHS) to extend the "Safer at Home" order until May 26. Backlash from legislators and the public was immediate, partisan, and spatially uneven. Republicans asserted Evers and the DHS exceeded their authority. By mid-April, Wisconsin was only averaging about 200 daily cases, with most of these cases found in the larger urban areas. Rural citizens across Wisconsin further agreed with Republican legislators that an extension of the "Safer at Home" orders were both unfair and causing economic hardships. After all, in the early stages Covid's spread was viewed by some as simply an urban phenomenon.

On May 13, 2020, the conservative majority of the Wisconsin Supreme Court backed a Republican-led lawsuit that Evers and the DHS exceeded their authority and struck down the state-wide order. The Court stated in its 4–3 decision that the "Safer and Home" order was "unlawful, invalid, and unenforceable." It went further, stating than any restrictions to battle Covid must be approved by the state legislature rule-making committee before they could be implemented. This ruling effectively left Evers powerless to dictate statewide mandates to combat covid (Hagemann 2020).

Though statewide bans were no longer allowed without the approval of Republicans in the legislature, local officials were not affected by the decision of the court. Resulting orders from local officials would take a highly partisan tone. Democratic-leaning counties like Dane and Milwaukee immediately issued orders in line with Evers's "Stay at Home" mandate. In counties that did not issue a mandate, cities were left to institute covid mitigation strategies. The result of these actions not only further cemented Covid as a partisan issue, but also reinforced already established partisan divides between rural and urban areas. By May 25, these divides would be further tested, but for a different reason.

On May 25, 2020, George Floyd was stopped by police outside a small market in Minneapolis, Minnesota. Police arrested Floyd on suspicion of using counterfeit $20 bills. During the arrest, Floyd was wrestled to the ground with his head eventually pinned to road by Minneapolis police officer Derrick Chauvin's knee. Video- and audio-captured by bystanders clearly showed the 46-year-old Floyd crying out for over nine minutes "I cannot breathe" and at one point, Floyd cried out for this mother. Even after the ambulance arrived, Minneapolis police officer Derrick Chauvin continued to kneel on Floyd's neck. Shortly after the ambulance appeared, Floyd died.

Protests ensued immediately after the death of Floyd. Yet, unlike earlier protests regarding police brutality, the murder of George Floyd sparked both national and international outrage. Though no one reason can explain why this murder increased consciousness regarding race and police brutality, it can be argued much can be contributed to Donald Trump and Covid. By the end June, nearly 22 percent of all Covid cases and 23 percent of deaths were among African Americans. One protester, Priscilla Borkor, stated "It's either COVID killing us, cops killing us, or the economy is killing us" (quoted in Altman 2020).

By June 1, the National Guard had been activated in 28 states and the District of Columbia. Though most protests were peaceful, some had turned violent. In Minneapolis, a local police station was destroyed, and many local businesses were burned and looted. President Trump did little to quell the protests and used inflammatory language toward protesters, calling them "thugs" and at one point stated "When the looting starts, the shooting starts" (quoted in Panetta 2020).

After the death of Floyd, race and police brutality took on a larger role in the 2020 presidential election. As protests intensified, common chants and signs said "Defund the Police." Though the meaning behind this phrase was easily misconstrued, Trump seized the opportunity to use it to his advantage. On June 4 he tweeted "The Radical Left Democrats new theme is 'Defund the Police … Remember that when you don't want Crime, especially against you and your family. This is where Sleepy Joe is being dragged by the socialists. I am the complete opposite, more money for Law Enforcement!'" (quoted in Allen and Hunnicutt 2020). The issue of defunding police hampered Biden and Democrats throughout the campaign as they struggled on how best to respond.

By the fourth of July, both protests and Covid cases began to subside. Yet Donald Trump refused to let either issue go. In two speeches, one in front of Mount Rushmore on July 3 and another at the White House on July 4, Trump warned the nation of the dangers associated with "the radical left." He stated, "We are now in the process of defeating the radical left, the Marxists, the anarchists, the agitators, the looters, and people who, in many instances, have absolutely no clue what they are doing." Trump also made the claim that 99 percent of Covid cases were "totally harmless" and used a favorite racist term, "kung flu," to mock the virus and place the blame on China (Reston 2020). At both speeches Trump and his guests refused to follow CDC guidelines of wearing masks or practicing social distancing. Joe Biden continued to make speeches and talked to the American public from his basement.

By the start of the virtual Democratic Convention on August 17, Covid cases had begun to rise once again. The hardest-hit states were Arizona and Texas. As cases spread, hospitals and morgues began to fill. Even staunch Texas Republican Governor Greg Abbot was forced to initiate a mask mandate in hopes to curb the spread of Covid. Democrats' decision to hold the convention virtually and strictly adhere to CDC guidelines was a clear attempt to set them apart from Trump and much of the Republican Party that continued to ignore and mock masks and social distancing. Yet, as Covid cases continued to increase, on August 23, the nation, and Wisconsin, were forced to reexamine police and race relations once again.

In the afternoon on August 23 in the city of Kenosha, Wisconsin, police were called to the scene of a domestic incident. A woman called 9-1-1 stating her boyfriend, Jacob Blake, was attempting to steal her car. Upon arrival, the police were informed Jacob Blake had a warrant for his arrest for a prior incident with the woman that had placed the 9-1-1 call. Though reports vary, Blake had an initial encounter with police as they attempted to apprehend them, but he was able to evade them and walked to his car. As he entered his car, a police officer, believing Blake had a knife, fired seven shots at Blake. Four bullets hit him in the back. Blake was rushed to the hospital and survived his wounds, but was paralyzed from the back down. With the entire situation caught on video, the backlash was immediate. Over the next few evenings protests engulfed Kenosha.

Kenosha is a city of nearly 100,000 people and is located 30 miles from Milwaukee and 60 miles from Chicago. The city was once an important automobile manufacturing center, but after the AMC plant closed in 1988, it has struggled to find a new identity. Located on the northern-most stop of Metra Commuter rail connecting Kenosha to Chicago's Loop, redevelopment in the wake of the AMC plant closing has been geared toward people taking the train to and from Chicago for work. The lower housing prices geared to commuters has increased the white population in the downtown areas that have traditionally been populated with larger proportions of African American. Outside of downtown Kenosha is surrounded by affluent white suburban communities. On August 25, a 17-year-old boy, Kyle Rittenhouse, left his shift as a lifeguard in Kenosha. Rittenhouse and his friend decided to clean up graffiti left from the previous evening's protests. The teenager received information that a local car dealership was looking for help to protect its business from protesters. Rittenhouse then grabbed an assault rifle and headed to the dealership (Danbeck and Jordan 2020). As protests continued that evening, Rittenhouse was involved in two separate incidents in which the teenager shot and killed two people and injured another. Rittenhouse was later apprehended, but the controversies surrounding both Blake and the 17-year-old further heightened the partisan divides of the 2020 election.

The troubles in Kenosha forced both presidential campaigns to respond. Trump scheduled a visit to Kenosha on September 1, but both Governor Evers and the mayor of Kenosha asked him not to attend. Trump showed up anyway, but never met with Blake or his family, though the visit allowed Trump to expand on his earlier calls for protecting police and stopping violent protestors from destroying property. Trump also could not resist commenting on Kyle Rittenhouse. The president liked a tweet that suggested Rittenhouse was model citizen and a reason to vote for Trump. The president also suggested that Rittenhouse was acting in self-defense (Wise 2020). Biden traveled to Wisconsin for the first time in the campaign, meeting with Blake's family and speaking with Blake on the telephone. Though many voters appreciated the visits from Biden and Trump, some voiced concerns both were simply exploiting the situation. Jamar Mayfield questioned of Biden "Is he just trying to get a vote?" (quoted in Alter 2020), while Tom Gram, whose camera business was destroyed in the riots, criticized Trump, saying "I think everything he does turns into a circus" (BBC 2020). Overall, the tepid response in Kenosha for both candidates was a microcosm of the difficulties all Americans were having in attempting to understand all that had happened during the summer of 2020.

The Covid-19 Surge

As summer gave way to fall, Covid continued to plague the election. In Wisconsin, Covid infections saw a slight increase in mid-July; by September 1, however, cases began to fall. Though the decline was seen as good news, the start of the school year had many nervous for what the fall would bring. In response, on July 30 Governor Evers issued a statewide emergency order requiring facemasks to be worn. Though either adherence to or disregard of the mandate only strengthened urban and rural divides. While many urban areas already had mask mandates in place, Evers's newest mandate only brought disdain in rural areas of Wisconsin. In an act of defiance, many county sheriffs across Wisconsin refused to enforce mask mandates. Washburn County Sheriff, Dennis Stuart, stated he would not enforce the order and that the order was a "government overreach and unconstitutional on many levels" (quoted in Kremer 2020). Schools also represented this divide. Urban school districts like Madison, Milwaukee, and Eau Claire started the school year either online or with a hybrid model requiring students to wear masks when they did attend school. Smaller school districts typically had either very limited requirements, such as requiring masks, or nothing at all.

By mid to late September, Covid was on the rise throughout the Badger State. The initial increase was found among college students aged 18 to 24. On September 1, there was, on average, nearly 1,000 new daily cases, by October 1, that number increased to around 2,000. Once thought of as simply an urban problem, Covid began infiltrating all areas of the state. Places like Green Bay, Appleton, and Oshkosh were hardest-hit as residents from urban and rural areas were filling ICU beds. By October 2, the Wisconsin Hospital Association showed that 87 percent of hospital beds in the eastern areas of the state were filled, mostly by Covid patients, and Belin Hospital, in Green Bay, was at 94 percent capacity (Rogan et al. 2020). The cities of Milwaukee, Racine, and Kenosha also saw increasing rates. As a result of the surge in the southern areas of the state, Governor Evers directed the National Guard to build a 530-bed field hospital in the Milwaukee County State Fair Grounds to help alleviate the stress on local health systems.

The increase in Covid activity throughout the state also had an impact on the presidential campaign. Though Trump was reduced to giving speeches in airplane hangars, he still drew thousands to his rallies. At all his rallies, there was no social distancing and few, if any, wore masks (Siders 2020). The lack of respect for Covid by Trump's followers at these rallies and of the lack of empathy by Trump himself led many to question whether these rallies should be held at all. On October 1, the mayor of La Crosse, Tim Kabat, became the first official to request Trump cancel his campaign rally there. Due to consistently high levels of Covid infections throughout the areas, Kabat said

> From a perspective of trying to slow the spread of coronavirus and trying to reduce our case numbers and get the challenges we're facing here

in La Crosse under control … I understand with a campaign season and a presidential election, all of these things will become political but we're trying to do right by our community.

(quoted in Beck and BeMiller 2020)

The rally was moved to an airport in Janesville, Wisconsin. A day later, on October 2, Trump again canceled a rally after the mayor of Green Bay made a similar request.

Though Trump had consistently mocked the severity of Covid, and the guidelines set by the CDC to halt the spread, not even the president could avoid the virus. On the evening of October 2, news broke that Donald Trump had been diagnosed with Covid. He and his wife both tested positive and cameras followed the president as he walked across the White House lawn to Marine One to be flown to Walter Reed Medical Center. In the next few days Trump was given experiential treatment and was nearly put on a ventilator himself. Covid had almost taken the president's life. Yet, in less than a week, Trump returned to the White House and though he was often questioned about the virus, he was surprisingly silent on the topic. It was unknown at the time what impact this sickness may have on the campaign, but it was clear it would not change Trump's view on the disease, on masks, or on social distancing.

By late October, Wisconsin was mired in its worst Covid surge to date. As stated earlier, the average number of new daily cases hovered around 2,000 on October 1, but that number tripled to over 6,000 by election day in November. Yet both nationally and in Wisconsin Republicans did little to stop the spread. Any attempts by Democratic Governor Tony Evers to discuss the topic either in person or in a special session of the legislature was either never accepted or gaveled out in a matter of seconds. Figure 6.6 shows the rate of daily Covid infections per 100,000 people for Wisconsin counties on November 2, 2020. This map clearly illustrates the eastern half of Wisconsin was hit the hardest and that surges would accelerate across the state in the weeks to come. Furthermore, rates were also highest in many of the rural counties in northern and western Wisconsin. The state's surge did not begin to decline until mid-November when new daily cases hovered around 8,000. Based on the surge of infections leading up to the election and the controversies revolving around Covid-19 itself in Wisconsin, the virus would be the defining issue of the 2020 election in the Badger State.

The November Election Results

By election night an eerie sense of déjà vu began settling in for Democrats across the nation. Much like Hillary Clinton in 2016, Joe Biden held a strong national lead in polling and in key battleground states, but the lead-up to 2020 was vastly different. Compared to Clinton, Biden's lead in the polls had been relatively steady since September. Though Republicans tried to make the Hunter Biden story something of an "October Surprise," much like the

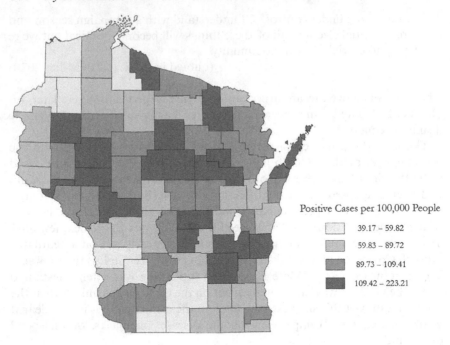

Positive Cases per 100,000 People

39.17 – 59.82

59.83 – 89.72

89.73 – 109.41

109.42 – 223.21

Figure 6.6 The 7 Day Average of Positive Covid-19 Cases per 100,000 Persons for Wisconsin Counties on November 4, 2020.

Sources: Map by author; data from Wisconsin State Legislature and the Wisconsin Department of Health Services

Comey letter in 2016, it had no staying power. In the weeks leading up to the election, however, the issue of ballots themselves played a pivotal role in the election. As was the case in Wisconsin in the April primary, absentee ballots became a truly partisan issue. Trump attempted to vilify them in an attempt to discredit the results, while Democrats clearly embraced them. Pundits, in the weeks leading up to November 3, had been attempting to warn the public of both "Blue and Red Mirages" on election night (Wasserman 2020). These refer to how states were counting both in-person and absentee ballots. Depending on vote counting procedures, one candidate may hold a huge advantage early on, only to see the large lead fade away as all votes are counted.

In Wisconsin, the chances that 2020 would be a carbon copy of 2016 were highly unlikely. For Trump to win the Badger State again, he would have had to replicate and improve upon his 2016 results. One reason this would be difficult for Trump was the lack of viable third-party candidates in 2020. In 2016, both Gary Johnson and Jill Stein had been on the ballot. Gary Johnson, the Libertarian, was a popular choice for college students due to his proclivity for the legalization of marijuana. He garnered over 120,000 votes, mainly in districts with college campuses. Jill Stein, as the Green Party candidate, also gained votes in similar areas. Though not all those voters would have voted for Clinton, it could be argued many would have done so if the third-party candidates had not been on the ballot. The failed attempt by Trump in 2020 to help Kanye West get on the ballot in Wisconsin was clearly a ploy to siphon away votes from Biden. In 2020, no third-party candidate would widely impact either candidate.

When polls closed in Wisconsin at 8 p.m., the state was quickly classified by all media outlets as "too close to call." Wisconsin law does not permit local election officials to count absentee ballots until the day of the election. According to the Wisconsin Election Commission, nearly 1.9 million absentee ballots were returned by November 3. The release of results throughout Wisconsin followed similar patterns to previous elections. The first results came in from mainly rural areas of the state and were mostly same-day votes. Trump's "Red Mirage" took shape as early returns showed Trump holding an early and commanding lead. As the night progressed, however, Trump's lead began fading as larger urban areas began reporting both same day and absentee results. By midnight, Dane County had declared its full results. The Democratic stronghold showed a high level of voter turnout coupled with extremely high support for Biden. The election was deadlocked, with only Milwaukee County yet to report results from the city of Milwaukee itself.

At 2:30 a.m. on November 4, Trump appeared on national TV, declaring he had won the election. Though no major media outlets made such a call, Trump also used the opportunity to question the authenticity of the election and the counting of ballots themselves. As Trump's "Red Mirages" slowly faded across the country, the wounded president, with no evidence, began attacking the counting of absentee ballots. In his words, the process was "a major fraud on our nation." Trump incoherently continued "We will win this and as far as I'm concerned, we already have won it ... We'll be going to the U.S. Supreme Court – we want all voting to stop. In fact, there is no more voting, just counting" (quoted in Associated Press 2020).

At around 3:40 a.m., Milwaukee County updated its election results. With the addition 69,000 absentee ballots from Milwaukee, Biden erased Trump's 31,000 lead (Slodysko 2020). Though Wisconsin was not declared for Biden until later in the day on November 4 by the Associated Press, Biden flipped Wisconsin by only 0.63 percent, or 20,682 votes. Scott Walker's 2017 law requiring candidates to pay for recounts if the total was between 1 and 0.5 percent hindered Trump's desire for a full statewide recount. In a show of desperation and political posturing, Trump eventually requested a recount for only Dane and Milwaukee counties, but in the end the recount only added to Biden's total votes. Coupled with victories in Pennsylvania, Michigan, Georgia, and Arizona, Joe Biden defeated Donald Trump.

Figure 6.7 shows both the county level and voting district results of the 2020 presidential election in Wisconsin. The 2020 election in Wisconsin can be summarized by a few key points. Biden's victory was aided by extremely high turnout in Dane County and wavering support for Trump in suburban areas around Milwaukee and the Fox River Valley. As polls did in 2016, they missed Trump's support among rural voters. The Republican was able to

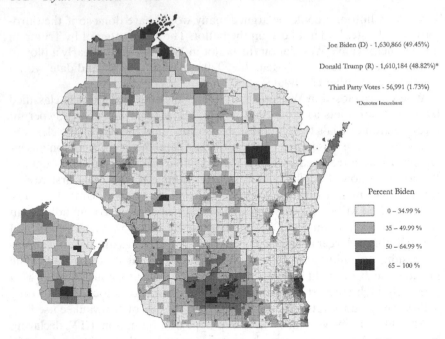

Joe Biden (D) - 1,630,866 (49.45%)

Donald Trump (R) - 1,610,184 (48.82%)*

Third Party Votes - 56,991 (1.73%)

*Denotes Incumbent

Percent Biden

0 – 34.99 %

35 – 49.99 %

50 – 64.99 %

65 – 100 %

Figure 6.7 Percent Joe Biden Vote for the 2020 Presidential Election at the County and Voting District Level.

Sources: Map by author; data from Wisconsin State Legislature and Wisconsin Election Commission

maintain much of his support in the many rural areas of the state. Trump's loss was also lessened by decreased turnout among African American voters in Milwaukee County. In contrast to the increased support among African-American voters in places like Detroit and Atlanta, African-American voters in Wisconsin were less inclined to vote in 2020. In the end, Biden's victory in Wisconsin can be attributed to high support of educated White voters in the many urban areas across the state. Unlike 2016, when many of these voters either did not vote or voted for a third-party candidate, their higher turnout was enough to exceed Trump's support in rural and traditional Republican strongholds across the state.

Figure 6.8 maps the percentage difference in Democratic vote between 2020 and 2016 for Wisconsin voting districts. Darker areas show where Biden overperformed Clinton and lighter areas show were Clinton overperformed Biden. There are a few areas were Biden underperformed Clinton. As was the case in the 2018 gubernatorial election and the 2020 primary, Democrats made inroads in some rural areas. Biden exceeded Clinton throughout almost all rural areas of the state, though urban turnout is what won Biden Wisconsin's ten electoral votes. The largest increases in Democratic support were in the many urban areas of Dane County. Clinton won Dane County with 70.3 percent of the vote in 2016. Aided by both contempt for Donald

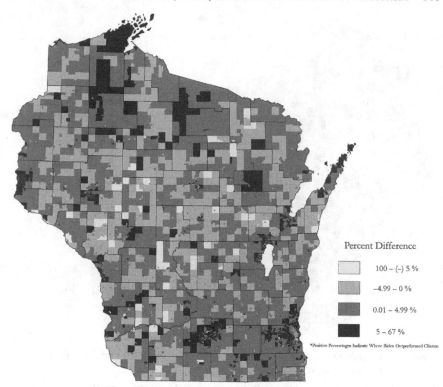

Figure 6.8 Percent Difference in Democratic Vote for the 2020 and 2016 Presidential Elections for Wisconsin at the Voting District Level.

Sources: Map by author; data from Wisconsin State Legislature and Wisconsin Election Commission

Trump and the lack of a third-party candidate, Biden increased that percentage to 75.5. Nearly 35,000 more people voted in Dane County in 2020 compared to 2016. Additionally, Biden gained over 42,000 more votes than Clinton. One of Clinton's bright spots in 2016 was in suburban areas around Milwaukee. That slow movement to the left was accelerated in 2020, where Biden saw greater support in Waukesha and Ozaukee Counties, both traditional Republican strongholds. Furthermore, Biden saw increased gains in the Fox River Valley near Appleton and Green Bay. These trends were seen in 2018 and continued into 2020. Compared to 2016, Biden's "poorest" showing was among African American voters in the city of Milwaukee. Total turnout between 2020 and 2016 only increased by 18,700 total votes, with Biden only increasing his margins compared to Clinton by 28,705 votes.

Figure 6.9 shows Trump's performance in 2020 compared to 2016. Based on this figure it is obvious Trump underperformed throughout much of the state compared to 2016. This figure clearly illustrates the rural and urban divide of Trump's support. Contrasting Walker in 2018 and Kelly in the 2020 Primary, Trump was able to maintain the enthusiasm in the rural areas.

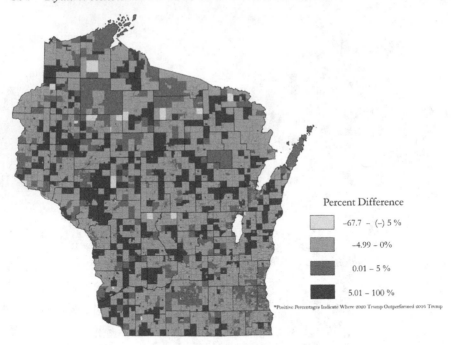

Figure 6.9 Percent Difference in Republican Vote for the 2020 and 2016 Presidential Elections for Wisconsin at the Voting District Level.

Sources: Map by author; data from Wisconsin State Legislature and Wisconsin Election Commission

Unlike 2016, Trump was unable to garner enough support in rural areas to overtake the increased Democratic turnout in large urban areas like Madison. The most likely explanation for Trump's loss were in suburban areas, especially in eastern Wisconsin, was Covid. The last Marquette Poll in Wisconsin showed Trump's approval rating regarding Covid dropped to a low 40 percent (51 percent in March) and his disapproval rate in handling the virus increased to a high of 58 percent. Though Evers also saw a decline in his response to Covid, his approval rate remained above 50 percent, at 52. Related to mask mandates, 64 percent of voters supported mask mandates (Conway 2020). Voters in Wisconsin undoubtably connected the state's Covid surge and Trump's poor handling of the pandemic at the ballot box on November 3. Trump did receive higher marks regarding police and protests. The Marquette Poll showed Trump's approval rating for handling protests at 30 percent in June. That figure increased to 40 percent before the election. Furthermore, voters also soured toward Black Lives Matter. Approval of the group dropped from 61 percent in June to 50 percent in late October (Conway 2020). It was clear the riots in Kenosha soured some voters, perhaps providing Trump some additional votes, but clearly not enough to sway the election in Wisconsin.

Conclusions

Perhaps no other state over the past decade has both impacted national politics as well as mirrored trends than Wisconsin. The election of Scott Walker and his Republican counterparts in the Tea Party wave of 2010 typified current urban and rural divides in American politics. In the years leading up to this watershed election, distain for urban areas had been growing in Wisconsin. Since 1990, growth in Wisconsin has been led by the large urban areas of Madison and Milwaukee. This concentration of population led many to believe attention and resources were unfairly distributed to this area of the state while forgetting struggling rural areas. Walker and his policies only highlighted these divides and drove wedges across the electorate. As the decade progressed, normally dependent blue rural voters faded into a sea of red. For these many reasons, the success of Donald Trump in 2016 in Wisconsin is perhaps not the surprise so many believed.

After the 2018 midterms it was clear the 2020 election would be a battle between urban and rural voters. The results of these elections provided Democrats a roadmap to victory in 2020. Before the midterms, it was unknown if Trump supporters would vote without Trump's name on the ballot. It was also unknown how energized Democrats would be. In Wisconsin, the answers to both questions were obvious. Rural voters would not demonstrate the same enthusiasm for Scott Walker as they did for Donald Trump and increased urban support in places like Madison could alone win an election. Furthermore, the cracks in Republican support in suburban areas seen in 2016 continued into 2018. To win in Wisconsin in 2020, Democrats simply had to expand turnout in cities to overturn any gains in rural areas and exploit the modest gains seen in suburban areas.

In the early months of the Democratic primary, it was unknown if any candidate could carry the momentum of 2018. With 27 candidates from which to choose, it was anyone's guess who would win. Though Democrats would not find a frontrunner until late March 2020, it was obvious Covid would be the defining issue of the campaign. The Wisconsin primary election would embody the controversies surrounding Covid and the 2020 election. As both parties dug in their heels, attitudes toward Covid took on clear partisan and spatial divides. Cities were at odds with rural areas in attempts to slow the spread of the virus and conduct safe elections through increased use of absentee ballots. Trump exacerbated this divide by often mocking the virus and claiming absentee ballots increased fraud to discredit the election results. As the primary ended and the summer gave way to fall, not only did the virus surge but so too did political divides. Even after Trump contracted Covid himself, it did little to dissuade his most ardent supporters to take the virus seriously.

Race was also a key issue in the 2020 election. The shooting of Breonna Taylor by police in March and the murder of George Floyd by police in Minneapolis in May both triggered hundreds of protests across the globe. The protests themselves took on a partisan divide after Trump suggested looters should be shot and when calls for defunding police departments grew.

These events allowed Trump to declare himself the "Law and Order President" and placed Democrats in a difficult position to respond. The issue became more complicated after the shooting of Jacob Blake by police in Kenosha and the shootings by Kyle Rittenhouse in protests that erupted thereafter. Both candidates used the tragedies to highlight the issues on the campaign trial.

It was clear Covid and race were the defining issues of this election, and that Wisconsin provided the lens to understand these through analyzing its election results. The many debates around these issues in the 2020 campaign not only highlighted urban and rural political divides, but also accelerated them. The political demagoguery of political actors like Donald Trump and Scott Walker continue to haunt the nation. The dangers of misinformation exposed in the 2020 election continue to divide America as lies about masks and Black Lives Matter protests have simply transformed and intensified into lies about vaccines and Critical Race Theory. Though these realities have resulted in rather static and predictable political landscapes, Wisconsin does demonstrate how closely divided Americans are.

References

Allen, J., and T. Hunnicutt. 2020. U.S. protesters call to "defund the police." What would that look like? *Reuters* (June 5). https://www.reuters.com/article/us-minneapolis-police-defunding-explaine/u-s-protesters-call-to-defund-the-police-what-would-that-look-like-idUSKBN23C2I9

Alter, C. 2020. Joe Biden visits Kenosha. *Time* (Sept. 3). https://time.com/5886059/joe-biden-kenosha-jacob-blake/

Altman, A. 2020. Why the killing of George Floyd sparked an American uprising. *Time.* https://time.com/5847967/george-floyd-protests-trump/

Associated Press. 2020. Donald Trump to supporters: "As far as I'm concerned, we already have won". *Boston.com* (Nov. 4). https://www.boston.com/news/politics/2020/11/04/donald-trump-addresses-supporters/

BBC. 2020. Jacob Blake: Trump visits Kenosha to back police after shooting. *BBC News* (Sept. 2). https://www.bbc.com/news/world-us-canada-53989076

Beck, M., and H. BeMiller. 2020. Trump moves rally to Janesville after La Crosse Mayor asks him to cancel; event in Green Bay still planned. *Gazette* (Oct. 1). https://www.greenbaypressgazette.com/story/news/2020/10/01/la-crosse-mayor-asks-trump-cancel-rally-covid-19-cases-surge-green-bay/5880327002/

Bice, D., and M. Spicuzza. 2018. With governor's race deadlocked, Milwaukee delivered for Evers with late absentee ballots. *Milwaukee Journal Sentinel* (Nov. 6). https://www.jsonline.com/story/news/politics/elections/2018/11/06/milwaukee-county-has-more-than-50-000-uncounted-absentee-ballots/1916187002/

Bonikowski, B. 2016. Three lessons of contemporary populism in Europe and the United States. *Brown Journal of World Affairs* 23 (1):9–23.

Campbell, C. 2015. Donald Trump is going to war with Scott Walker after being called 'DumbDumb' by one of his supporters. *Business Insider* (July 27). https://www.businessinsider.com/donald-trump-scott-walker-dumbdumb-2015-7

Cillizza, C. 2019. How Donald Trump destroyed Scott Walker's presidential chances. *The Washington Post.* https://www.washingtonpost.com/news/the-fix/wp/2015/09/21/how-donald-trump-destroyed-scott-walkers-presidential-chances/

Conway, K. 2020. New Marquette Law School poll finds Biden lead over Trump stable at five percentage points. News Center, Marquette University. https://www.marquette.edu/news-center/2020/new-marquette-law-school-poll-finds-biden-lead-over-trump-stable-at-five-percentage-points.php

Corasaniti, N., and S. Saul. 2020. "Your health or the right to vote": A battle in Wisconsin as its primary nears. *New York Times.* https://www.nytimes.com/2020/03/27/us/politics/wisconsin-primary-coronavirus.html

Cramer, K. 2016. *The Politics of Resentment: Rural Consciousness in Wisconsin and the Rise of Scott Walker.* Chicago: University of Chicago Press.

Danbeck, J., and B. Jordan. 2020. Attorneys representing Kyle Rittenhouse say he was wrongfully charged after "acting in self-defense." *WTMJ-TV* (Aug. 28). https://www.tmj4.com/news/local-news/attorneys-representing-kyle-rittenhouse-say-he-was-wrongfully-charged-after-acting-in-self-defense

Davey, M. 2018. Tony Evers wins Wisconsin governor's race; Scott Walker concedes. *The New York Times* (Nov. 7). https://www.nytimes.com/2018/11/07/us/elections-wisconsin-governor-evers-walker.html

Gass, N., and K. Glueck. 2016. Scott Walker endorses Ted Cruz. *Politico* (March 29). https://www.politico.com/story/2016/03/scott-walker-endorses-ted-cruz-221327

Hagemann, H. 2020. Wisconsin Supreme Court overturns the State's "Stay-at-Home Orders." *NPR.* https://www.npr.org/sections/coronavirus-live-updates/2020/05/13/855782006/wisconsin-supreme-court-overturns-the-states-stay-at-home-order

Kremer, R. 2020. Evers' administration says statewide mask mandate can be enforced despite sheriffs' opposition. *Wisconsin Public Radio* (Aug. 5). https://www.wpr.org/evers-administration-says-statewide-mask-mandate-can-be-enforced-despite-sheriffs-opposition

Kremer, R., and S. Johnson. 2018. Dane, Waukesha counties tell tale of Wisconsin's 2018 election for governor. *Wisconsin Public Radio* (Nov. 7). https://www.wpr.org/dane-waukesha-counties-tell-tale-wisconsins-2018-election-governor

Lafond, N. 2018. Walker warns of "#bluewave" coming to WI after Dem wins state high court race. *Talking Points Memo* (April 4). https://talkingpointsmemo.com/livewire/walker-warns-blue-wave

Levin, B., and H. Busis. 2018. Trump's "incredible" Foxconn deal turns out to be another massive con job. *Vanity Fair* (Nov. 5). https://www.vanityfair.com/news/2018/11/donald-trump-foxconn-scam

Linnane, C. 2017. Foxconn's history of broken promises casts a shadow on Wisconsin news. *MarketWatch* (July 31). https://www.marketwatch.com/story/foxconns-history-of-broken-promises-casts-a-shadow-on-wisconsin-news-2017-07-27

Marley, P., and K. Crowe. 2018. Scott Walker raises $7.7 million this fall; Tony Evers takes in $6 million. *Milwaukee Journal Sentinel* (Oct. 29). https://www.jsonline.com/story/news/politics/elections/2018/10/29/scott-walker-tony-evers-report-fundraising-totals-final-campaign-stretch/1806651002/

Murphy, C. 2018. Wisconsin's $4.1 billion Foxconn boondoggle. *The Verge* (Oct 29). https://www.theverge.com/2018/10/29/18027032/foxconn-wisconsin-plant-jobs-deal-subsidy-governor-scott-walker

Panetta, G. 2020. Trump claims his "when the looting starts, the shooting starts" remarks weren't a call to violence but instead a "fact". *Business Insider* (May 29). https://www.businessinsider.com/trump-defends-his-when-the-looting-starts-the-shooting-starts-tweet-2020-5

Paquette, D., and T. Frankel. 2019. Foxconn announces new factory in Wisconsin in much-needed win for Trump and Scott Walker. *Washington Post* (July 26). https://www.washingtonpost.com/news/wonk/wp/2017/07/26/foxconn-to-announce-new-factory-in-wisconsin-in-much-needed-win-for-trump-and-scott-walker/

Queally, J. 2020. Trump admits "you'd never have a Republican elected in this country again" if voting access expanded. *Salon* (March 31). https://www.salon.com/2020/03/31/trump-admits-youd-never-have-a-republican-elected-in-this-country-again-if-voting-access-expanded_partner/

Reston, M. 2020. Trump uses July 4th address to put forward a dangerously misleading claim. *CNN* (July 5). https://www.cnn.com/2020/07/05/politics/donald-trump-july-4-coronavirus/index.html

Richmond, T. 2020. Judge won't delay Wisconsin election but extends voting. *AP News* (April 2). https://apnews.com/article/e36c3adc0c7474014f3e7ab566071303

Rogan, A., C. Thayer, and S. Jones. 2020. COVID hospitalizations on the rise in Southeastern Wisconsin, but situation is much worse further north. *Journal Times* (Oct. 2). https://journaltimes.com/news/local/covid-hospitalizations-on-the-rise-in-southeastern-wisconsin-but-situation-is-much-worse-further-north/article_d5296a66-cebf-531e-aeaa-b86237cf55fa.html

Root, D. 2020. Wisconsin primary shows why states must prepare their elections for the Coronavirus. *Center for American Progress*. https://www.americanprogress.org/issues/democracy/news/2020/04/27/484013/wisconsin-primary-shows-states-must-prepare-elections-coronavirus/

Shafer, J. 2015. Donald Trump, American Demagogue. *Politico* (Aug. 10). https://www.politico.com/magazine/story/2015/08/dont-write-trumps-obit-yet-121232/

Siders, D. 2020. Nine coronavirus cases tied to Trump Minnesota rally. *Politico* (Oct. 9). https://www.politico.com/news/2020/10/09/trump-minnesota-rally-coronavirus-cases-428425

Slodysko, B. 2020. Why AP called Wisconsin for Biden. *AP News* (Nov. 4). https://apnews.com/article/ap-explains-wisconsin-joe-biden-636a771c35314b13a5e33cb19092f9d5

Smith, M., and M. Davey. 2018. Wisconsin's Scott Walker signs bills stripping powers from incoming governor. *New York Times* (Dec. 14). https://www.nytimes.com/2018/12/14/us/wisconsin-governor-scott-walker.html

Sommerhauser, M. 2017. Tony Evers campaign ad hits Scott Walker on Foxconn deal. *La Crosse Tribune* (Aug. 29). https://lacrossetribune.com/news/article_f832ebb9-5d7e-5e4f-ac5a-2d83993dd354.html

Wasserman, D. 2020. Beware the "blue mirage" and the "red mirage" on election night. *NBCNews.com* (Nov. 3). https://www.nbcnews.com/politics/2020-election/beware-blue-mirage-red-mirage-election-night-n1245925

White, L. 2020. Polls missed the mark in 2016. but experts say things are different in 2020. *Wisconsin Public Radio* (Oct. 19). https://www.wpr.org/polls-missed-mark-2016-experts-say-things-are-different-2020

Wise, A. 2020. Trump defends Kenosha shooting suspect. *NPR* (Aug. 31). https://www.npr.org/sections/live-updates-protests-for-racial-justice/2020/08/31/908137377/trump-defends-kenosha-shooting-suspect

7 Political Ramifications of the Jacob Blake Shooting in Kenosha, Wisconsin on the 2020 Presidential Election

Kenneth French and Ryan Weichelt

Since 2015, over 5,800 people in the U.S. have been fatally shot by police while in the line of duty. Based on the total numbers of shootings, White victims accounted for 2,962 deaths, Black victims 1,554, Hispanic 1,084, and 244 fell into an "Other" category. When normalized by total population for each group, Black Americans are shot and killed by police at a rate more than double compared to white Americans (*Washington Post* 2021). In this same period, shootings and police brutality captured on video also increased. In 2018 the *New York Times* collected and posted over 30 videos capturing examples of police brutality and the murder of non-white victims. In some instances, the videos had little, if any influence, while others, such as the shooting of Michael Brown in 2014, sparked nights of protest. No matter the result of any such videos, they provide a consistent reminder of race relations in the United States.

Though the 2020 presidential election was mostly about Covid-19, numerous instances of police brutality during the campaign consistently reminded the nation and the candidates of the seriousness of the topic. On March 13, 2020, non-uniform police agents in Louisville, Kentucky served a "no knock" warrant for Kenneth Walker at the apartment of his girlfriend, Breonna Taylor. As police broke open the door, a gun fight ensued. Walker claimed the police never announced their presence and that, thinking they were intruders, he shot an officer in the leg. In response, police officers blindly fired 32 shots into the apartment. Unarmed and caught in the crossfire, Taylor was shot six times and died. Though protests lasted over 75 days, Taylor was not the last victim of police violence in 2020.

On May 23, George Floyd was flagged down by police and questioned for the potential use of counterfeit $20 bills. Shortly after police stopped him, though Floyd had not resisted arrest, Floyd was handcuffed and placed in a chokehold with his head pinned to the concrete by the knee of officer Derek Chauvin. Video of the incident clearly identified Floyd's voice as he repeatedly screamed "I can't breathe." For over nine minutes Floyd was slowly suffocated. As the ambulance arrived, a lifeless Floyd remained under the knee of Chauvin. Floyd was pronounced dead moments later. Reaction in Minneapolis was immediate and sent shock waves across the nation and the 2020 presidential election campaign.

DOI: 10.4324/9781003260837-7

George Floyd Protests June 2020
and 2020 Presidential Election Results

Figure 7.1 Location of George Floyd and Black Lives Matter Protests – May 26 to
June 28.

Source of data: Alex Smith.

After the murder of George Floyd, protests popped up across the United
States and the world. By the end of June, over 3,294 protests had occurred
across the United States in cities and small towns alike (Smith 2020). Figure 7.1
illustrates the locations of George Floyd and Black Lives Matter protests
from May 26 to June 28. Previous incidents, even the recent one involving
Breonna Taylor, did not capture the attention of the nation like Floyd's death.
Though most protests would be peaceful, riots and looting in Minneapolis
and other locations across the United States provided Donald Trump the
opening he needed to drive a wedge between voters and protestors.

In the months that followed, Trump attempted to exploit this situation
for his own political gain. The consistent message of condemning protestors
as violent actors played well to Trump's predominantly White base. By mid-
August, polls across the nation began showing decreased support for protestors and the Black Lives Matter movement, especially among White
respondents.

On August 23, Jacob Blake was violently shot seven times in the back by
police officers. Though Blake survived, the video that surfaced reignited the
debate around race and police brutality. In the protests that followed, a
17-year-old teenager, Kyle Rittenhouse, grabbed an assault rifle and walked
to downtown Kenosha, heeding the call to help protect local businesses. In
the chaos, Rittenhouse shot three individuals, killing two. The disorder of
those days in late August would change the face of the 2020 election. Both
candidates were forced to respond and both candidates traveled to Wisconsin
in the days after Blake was shot. There was no doubt after the summer of
unrest that the chaotic events in Kenosha would further impact the election
at both the national and local scales.

The Jacob Blake shooting, the subsequent protests, and Kyle Rittenhouse
all had an impact on the conduct of the 2020 election. Though national

results illustrated distinct rural, suburban, and urban divides, analysis of the election results in Kenosha, Wisconsin will provide further insight into the potential impact the events of late August had on the local political landscape. Because the entire situation cut deep across both political and cultural lines, Kenosha provides an excellent canvas to analyze such impacts. The entirety of Kenosha County is a mixture of wealthy White suburban communities surrounding the urban core of the city of Kenosha populated with higher proportion of African-American citizens. Therefore, this chapter will explore whether the entire Blake situation created a distinct electoral landscape in Kenosha, with specific emphasis on suburban and urban voting patterns.

Kenosha's Economy and Demographics

The city of Kenosha is located in the southeastern-most part of Wisconsin on the banks of Lake Michigan. There are 99,944 people living in the city of Kenosha and 169,561 people in the wider Kenosha County (U.S. Census 2019). The city is located in the middle of the Milwaukee-Chicago conurbation and Kenosha County is the northern-most place in the Chicago Metropolitan Statistical Area. Kenosha's economic base, similar to similar-sized cities in the Midwest, was tied to manufacturing and these industries built a strong working class protected by powerful unions. However, the city continues to lose its industrial base – the Chrysler Engine Plant closed down in 2010 to no fanfare (Barrett and Taschler 2010) and its buildings have since been torn down. Kenosha continues to slowly transition into a bedroom community as it is common for commuters to work in either Chicago or Milwaukee. The main businesses in the Kenosha area include Jockey International (apparel), Snap-on (tool manufacturer), an Amazon warehouse located along Interstate-94, and stores in the Pleasant Prairie Premium Outlets. Today, the area is a mix of working-class residents in the heart of Kenosha and former Illinoisans in the suburbs taking advantage of the lower taxes and cheaper housing. Overall, the city of Kenosha has a per capita income of $28,680 and 16.0 percent of its population lives in poverty, a figure almost 3.0 percent higher than the national percentage (U.S. Census 2019).

Given the working-class background of the city of Kenosha, it is not surprising that only 24.4 percent of the population over 25 years of age has a bachelor's degree or higher (compared to 36.0 percent for the nation as a whole) (U.S. Census 2019). Kenosha County has one public four-year university and one private university, the University of Wisconsin-Parkside and Carthage College, respectively. UW-Parkside is the most racially diverse campus in the UW-System and is now recognized as an Emerging Hispanic Serving Institution (UWP 2021). Carthage College is selective in its admissions process; in the most recent available statistics, 7,071 high school students applied for the 727 positions in the freshman class (Carthage 2021). Educational quality in the public school system is not that high, as the three

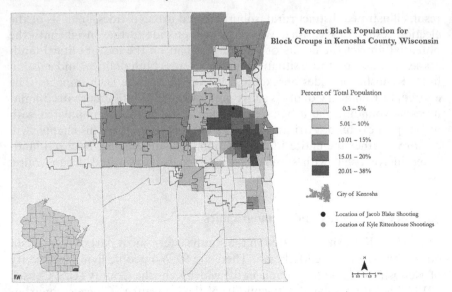

Figure 7.2 Percent African American per Block Group in Kenosha County.

Source: United States Census.

public high schools (Bradford, Indian Trails, and Tremper) have ratings of 3 out of 10 or below (Great Schools 2021).

The racial breakdown of the city of Kenosha consists of 66.1 percent White, non-Hispanic; 11.5 percent African American; 2.0 percent Asian American; 0.7 percent Native American; and 17.8 percent Hispanic/Latinx (U.S. Census 2019). The highest percentage of African Americans in Kenosha live in the older, poorer neighborhoods just west of the lakefront downtown (Figure 7.2). There are few African Americans living in the suburban area south (Pleasant Prairie) and west (Somers) of Kenosha's city limits. The Hispanic/Latinx populations also live in the neighborhoods just west of downtown Kenosha. The highest proportion of White, non-Hispanic populations live along Lake Michigan (north and south of downtown) and in neighborhoods further away from the center of the city.

The Jacob Blake Shooting

On August 23, 2020, Jacob Blake was shot seven times in the back for resisting arrest by a Kenosha police officer after a domestic incident call (Booker and Bowman 2020). This shooting of yet another African American by the police was recorded by onlookers and went viral, sparking social unrest in Kenosha and amplifying the Black Lives Matter protests of 2020. The police and Blue Lives Matter counter protesters believe the shooting was justifiable since Blake was in possession of a knife and the officer was in fear for his life. The shooting of Jacob Blake and subsequent social unrest became a political hot topic.

Jacob Blake was issued a warrant for his arrest in July 2020 for third-degree sexual assault, disorderly conduct, and criminal trespass (Morales 2021). While Blake and his kids were at a party on August 23, 2020, a woman called 911 to complain about Blake's presence at the party. The woman who had called 911 was the accuser whose actions led to the initial warrant for Jacob Blake. Blake's attorney, Ben Crump, described him as a peacemaker trying to break up a disturbance between two women (Eligon et al. 2020). As police arrived, Blake was initially tasered during a scuffle, but he eventually made his way to his vehicle. As Black entered his vehicle, with his children in the back, Blake was shot multiple times by Kenosha Police Officer Rusten Sheskey (Petras and Padilla 2020). In the days after the shooting, Kenosha police issued a statement that Officer Sheskey shot Blake out of fear he was reaching for a knife with the intent to stab him. The shooting on the 2800 block of 40th Street (Figure 7.2) left Mr. Blake paralyzed from the waist down and with additional injuries to his arm, kidney, and liver (McLaughlin et al. 2020). In accordance with the terms of his arrest warrant, the paralyzed Mr. Blake was handcuffed to his hospital bed as he recovered; he was later unhandcuffed following a public backlash. Months after this shooting incident, the sexual assault charge against Mr. Blake was dropped and he pled guilty to disorderly conduct.

The combination of already highlighted tensions regarding police brutality from the earlier Taylor and Floyd murders and aided by videos of Blake's shooting circulating on social media created instant outrage in Kenosha and the rest of the country. Nationally, the protests over what happened to Jacob Blake included the Milwaukee Bucks boycotting their playoff game and several sporting events cancelled on August 26 (Cohen 2020). Coming on top of the killing of George Floyd, Breonna Taylor, and Ahmaud Arbery, the shooting of Blake led to more Black Lives Matter marches in 2020.

The Black Lives Matter movement was begun in 2013 by Opal Tometi, Patrisse Cullors, and Alicia Garza in response to the acquittal of George Zimmerman over the killing of Trayvon Martin in Sandford, Florida (Lebron 2017). Over the years, the Black Lives Matter movement would encapsulate the frustration experienced by many non-White groups regarding police brutality. In the days after Blake's shooting, peaceful Black Lives Matters protests during the day were eventually overshadowed by nighttime violent protests. On the first night of protests, the police used garbage trucks on Sheridan Road to block vehicle access to downtown Kenosha. The initial night of protests included fire damage to a furniture supply store, a car repair business and the Department of Corrections field office (Rose et al. 2020). In addition to damage directed toward businesses, cars and garbage trucks were set on fire. As nightly protests continued, Governor Tony Evers (D) called on the National Guard to support local officials on August 24 and an 8 p.m. curfew was put in place by Mayor John Antaramian (Booker and Bowman 2020).

The protests were located in two parts of the city: the downtown and the Uptown neighborhood. In total, several Kenosha businesses and buildings were damaged over the duration of the protests (Figure 7.3), ranging from

Figure 7.3 Protest Damage from August 23–September 1, 2020 in Kenosha, Wisconsin.

Sources: Wisconsin State Cartographer's Office and the Kenosha Area Business Alliance.

broken windows to complete destruction. Several downtown buildings east of the main thoroughfare of Sheridan Road saw the majority of damage – consequently, these areas were also the most heavily protected by law enforcement. Also, several buildings along 60th Street, just outside of Kenosha's downtown, were also the target of protestors. However, the most buildings burnt to the ground were located along 22nd Ave., in the Uptown neighborhood that is west of downtown. The Uptown neighborhood represents high levels of poverty and had little, if any law enforcement presence during the protests. Seeing a political opportunity to campaign in a toss-up state, both 2020 presidential candidates visited Kenosha in the midst of the pandemic. President Donald Trump, in front of a burned-down camera shop on September 1, 2020, called the riots "domestic terror" and vowed to use federal funds to rebuild Kenosha (Miller and Lemire 2020). The owner of the camera shop, Tom Gram, criticized President Trump for using his store for political gain as the president stood next to the former owner of the shop (Jordan 2020). The Democratic candidate, Joe Biden, visited Kenosha on September 3, 2020, and talked to African American leaders at Grace Lutheran Church (Barrow et al. 2020). In the end, it is estimated that damage caused by the Jacob Blake protests in Kenosha topped $50 million (Flores 2020).

On January 5, 2021, the Kenosha County District Attorney (DA) Michael Graveley did not bring charges against Officer Sheskey for shooting Mr. Blake. The DA concluded that Jacob Blake was in possession of a knife and that Officer Sheskey stated he "feared Jacob Blake was going to stab him with the knife" (Richmond and Tarm 2021). Jacob Blake filed a Civil Rights lawsuit against Officer Sheskey, alleging the use of excessive force (Treisman 2021). The electoral ramifications did not end with a police officer shooting an African American man resisting arrest, as another incident during the nightly protests ignited yet another political firestorm.

The Kyle Rittenhouse Killings

On the night of August 25, 2020, 17-year-old Kyle Rittenhouse from Antioch, Illinois, traveled to Kenosha and shot three protestors – 26-year-old Anthony Huber and 36-year-old Joseph Rosenbaum, who both died where they were shot, and 26-year-old Gaige Grosskreutz, who survived (Johnson et al. 2020). Rittenhouse and his friends had answered a call for armed protestors via a Facebook post by a local militia group, the Kenosha Guard, to counter the protestors upset with the shooting of Jacob Blake (Timberg 2020). The goal of the Kenosha Guard was to weaponize citizens to protect Kenosha buildings and businesses from the rioters. Kyle Rittenhouse claimed that he was part of a group that was asked by a Kenosha car dealer to guard his property, but the car dealer denied asking for help (Vielmetti 2020). Rittenhouse stated that he shot the protestors in self-defense, because he was in fear for his life. He was arrested back home and charged with first-degree homicide, first-degree intentional homicide, attempted first-degree intentional homicide, two counts of recklessly endangering safety, and possession of a dangerous weapon while under 18 years of age (Treisman 2020). The shootings by Rittenhouse and his eventual arrest immediately took on clear partisan support. Many staunch conservatives proclaimed the teenager a hero for protecting businesses, while many liberals viewed Rittenhouse as a vigilante who shot innocent protestors.

The shootings took place outside of the downtown area along Sheridan Road, south of 60th Street (Figure 7.3). It is alleged that Joseph Rosenbaum verbally attacked Mr. Rittenhouse at the car dealership parking lot on the corner of 63rd Street and Sheridan Road. After shooting Rosenbaum multiple times, Rittenhouse fled north on Sheridan to avoid protestors who witnessed the shooting. Two protestors, Anthony Huber and Gaige Grosskreutz, allegedly tried to stop Rittenhouse, with Huber using his skateboard to knock the teenager to the ground. In the melee, Rittenhouse shot Huber in the midsection and then shot Grosskreutz in the arm (Maxouris 2020). Videos released after these shootings show Rittenhouse raising his hands in the air and then walking by oncoming police officers. Though multiple peopled indicated to police Rittenhouse was the shooter, the 17-year-old left Kenosha and returned to his mother's house in Illinois (Willis et al. 2020).

The juxtaposition between the Jacob Blake shooting and the Kyle Rittenhouse killings became an immediate political talking point. Many conservatives defended Rittenhouse and donated money to the "Fight Back" organization to pay his $2 million bail (Guarino 2020). He was seen by many as protecting Kenosha businesses and defending himself against violent rioters. Rittenhouse would also become a hero for the Alt-Right. On January 5, 2021, photographs emerged of Rittenhouse drinking in a Wisconsin bar flashing the "OK" White Power symbol and being serenaded by a group of men singing the Proud Boys' anthem (Wong 2021). Conversely, many Black Lives Matter protestors wondered if Rittenhouse would have been treated differently by the police if he were African American. In response to how Rittenhouse was treated, Jacob Blake said:

> That was like a kick right in the you know what. I was angry, I was furious, and I felt like I had every right to be. For the reasons they said they shot me, they had every reason to shoot him, but they didn't. Honestly, if his skin color was different, and I'm not prejudiced or a racist, he probably would have been labeled a terrorist.
>
> (Jimenez 2021)

The Blake shooting and Rittenhouse killings became a political Rorschach test. How voters viewed these violent incidents correlated well with political ideology and most likely influenced some voters in the 2020 presidential election. This was certainly the case in Kenosha, Wisconsin. Rittenhouse was acquitted by a jury in November, 2021.

2020 Election Results in Kenosha County

Wisconsin is a truly purple state. Before 2016, Wisconsin had been a Democratic stronghold. The Badger State had not elected a Republican candidate since Ronald Reagan in 1984, though most elections after 2000 have produced close results. After the 2012 election, patterns at the local level transformed into distinctive rural, suburban, and urban divides. Though rural areas provided Barack Obama consistent support in 2008 and 2012, the success of Tea Party candidates in 2010 highlighted cracks among rural Democrats, and, by 2014, rural areas completely transformed into the most ardent GOP voters. Suburban areas have seen slow, but have seen consistent declines in Republican support since 2016. Urban areas across the state provide Democrats the lion's share of total votes in any given election.

Though Kenosha County has few rural areas, the county can be described as suburban with a distinct urban core. The city of Kenosha is a consistently Democratic stronghold, but is surrounded by Republican suburban areas to the south and east of the city. These electoral landscapes have produced several close elections and varied election results since 2010. In 2010, Kenosha voters provided Republican Scott Walker a three-point victory, but in two years during the recall election, Democrat Tom Barrett won the county by 1.2 percent. In 2014, voters then again supported Walker by less than 2 percent, but then voted for Democrat Tony Evers by over 3 percent in 2018. At the presidential level, Kenosha voters heavily supported Obama in 2008 and 2012, but 2012 saw decreased support compared to 2008. In 2016, Trump won the county by only 0.31 percent of the vote, but in 2020 his margin of victory increased to 3.13 percent. Similar increases from 2016 to 2020 for Trump were seen in the counties surrounding Kenosha, bucking the trends of other suburban counties across the remaining areas of Wisconsin that saw either minimal increases or decline in support for Trump in 2020.

The fact that Trump saw increased support in the immediate areas surrounding Kenosha is due in large part to the Jacob Blake shooting, protests, Kyle Rittenhouse, and Covid-19. Trump was no stranger to using the topic of riots to further his campaign. The days after the murder of George Floyd, Trump famously emerged from the White House, Bible in hand, and made his way to St. Francis Church only a few blocks away. Minutes before Trump took his photo in front of the church, police cleared protestors with tear gas and riot gear to allow the president safe passage. The day after the infamous picture was taken, Trump retweeted a line "when the looting starts, the shooting starts" in response to the protests in Minneapolis. Though Trump tried to backtrack his claim, in the days that followed he entrenched his message, proclaiming further he was "the Law-and-Order President." This rhetoric was further exacerbated as protestors began flying banners and signs to "defund the police." Trump reveled in the chaos, while Joe Biden and Democrats struggled to respond.

A June 10 *USA Today*/Ipsos (2020) poll found Americans were split on whether law and order or the right to protest was more important. Though at the time of the poll, a strong majority (65 percent) supported the George Floyd protests occurring across the country, and 54 percent believed they were mostly peaceful. Yet Republicans were much more likely (54 percent) to say the protests were violent compared to Democrats, who overwhelming (73 percent) believed the protests were mostly peaceful. In relation to police limiting protests, over half of White and Republican respondents (51 percent and 69 percent, respectively) were more likely to support police action to quell protests, compared to African Americans and Democrats (67 percent and 65 percent, respectively) who believed the right to protest was more important. In nearly twin events on July 3 in front of Mount Rushmore and on July 4 on the lawn of the White House, Trump spent large portions of his speeches supporting law enforcement and vowing to defeat looters and agitators (CNBC 2020). As the summer progressed, Trump consistently pushed this narrative in almost all his campaign speeches.

By mid-August, though national polls were not as numerous on the topic, polls across the nation were showing decreased support for protests and the Black Lives Matter movement, with the greatest decline in support among White voters compared to similar polls taken in early June (Blake 2020). The most striking poll was by Marquette University, conducted in Wisconsin in July and August. Though results were released after the Jacob Blake shooting, questions were all conducted before the incident. The Marquette Poll replicated questions about protests and Black Lives Matter in both the June and the July and August polls. Results showed support dropped from 63 percent in June to 48 percent by August. The largest decrease was seen among White respondents, which dropped from 59 percent in favor and 38 percent opposed to 45 percent in favor to 51 percent opposed. Results among non-White respondents showed broad support and was consistent for both polls (Franklin 2021). Yet, while backing dipped among Whites, support for both Trump and Biden remained generally unchanged in national polls. The shooting of Jacob Blake on the afternoon of August 23 once again brought police shootings and protests back to the forefront in the campaign.

After the shooting of Blake, the subsequent protests, and the shootings by Kyle Rittenhouse, Kenosha, Wisconsin vaulted into the national spotlight, forcing both candidates to respond. Joe Biden took a more cautious approach reaching out to Blake's family, eventually traveling to Kenosha, meeting with his mother, and talking to Jacob Blake himself on the phone. Donald Trump took the opposite approach. Though he did attempt to call Blake's mother (who "missed" the call), the president issued no statement about the shooting and focused mainly on the protests that immediately ensued. Trump tweeted in the days after the incident he would "NOT stand for looting, arson, violence, and lawlessness on American Streets." An additional tweet continued, "TODAY, I will be sending federal law enforcement and the National Guard to Kenosha, WI to restore LAW and ORDER!" (Liptak 2020).

Further complicating the situation was Kyle Rittenhouse. Ten days after the shooting of Blake, Donald Trump and Attorney General, William Barr, announced they would be visiting Kenosha to survey the damage after nights of protests. Both Democratic governor Tony Evers and Kenosha mayor John Antaramian asked the president to delay his visit. Trump did not. In the days after Rittenhouse shot three men, killing two, Trump suggested at a White House Press Conference that the teenager was simply defending himself. Trump said "And that was an interesting situation. You saw the same tape as I saw. And he was trying to get away from them, I guess; it looks like. And he fell, and then they very violently attacked him" (McCarthy 2020). On October 1, an internal document showed Trump officials were told to make sympathetic comments about Kyle Rittenhouse in the days after the shootings. Though agencies never expressed any sort of sympathetic comments, the White House continued to push the narrative Trump expressed at the earlier press conference (Ainsley 2020).

As the election drew near, citizens in Wisconsin and Kenosha were attempting to ascertain what happened in the days after the Blake shooting. Trump and Biden represented two completely different sides of the argument in relation to protests and police violence. Biden struggled to ignore the "Defund the Police" chants often accompanying the summer protests. Though he made it very clear he supported the police, the court of public opinion often tied Democrats and Biden to both Black Lives Matter and "Defund the Police." Trump made it very clear he was the "President of Law and Order" and did not condone the violence and rioting that occurred throughout much of 2020. Marquette polls from June through October 4 all showed a steady decline in both approval and views of favorability for protests and Black Lives Matter. Though non-White support remained high, White respondents reported the greatest decline in support. The last Marquette poll conducted before the election, from October 21 to October 25, indicated a slight increase in support for protests compared to the most recent poll conducted in early October. Approval rates increased from 46 percent to 50 percent and favorability ratings saw slight increases as well. Though support of protests increased, so did approval of Trump's handling of the protests. Only 30 percent of respondents in June approved of Trump's handling of the protests. That number increased to 40 percent in late October and his disapproval rate for handling the protests dropped only 4 percentage points (Conway 2020).

By election night, Joe Biden led polls in Wisconsin by an average of five points, though as votes were slowly tabulated it became clear Wisconsin was up for grabs. Trump held an early lead as same-day voting was counted, but as the night progressed and both absentee ballots and larger urban areas released results, Biden quickly cut into Trump's lead. By 3 a.m. on Wednesday, November 4, Milwaukee County tallied all its votes and Joe Biden held a 20,000-vote lead, eventually netting Wisconsin's 10 Electoral College votes. Traditional election patterns held true to form across Wisconsin, with rural areas heavily supporting Trump, urban areas strongly supporting Biden, and suburban leaning Trump, but with decreased support compared to 2016.

Ultimately, high voter turnout in Dane County provided Biden the necessary support he needed, but the events of 2020 regarding race and police would weigh heavily on southeastern Wisconsin, especially Kenosha.

The 2020 election results in Kenosha County showed Donald Trump defeated Joe Biden. Though Trump also won Kenosha County in 2016, he saw a three percent gain in vote in 2020. Similar patterns were seen in Racine County to the north and Walworth County to the west. In total, Kenosha County saw an increase of 12,343 total votes compared to 2016. Figure 7.4 shows the increase total votes the voting district level. The largest total increase in votes was found in the suburban areas on the fringes of the city of Kenosha. The village of Pleasant Prairie to the south of Kenosha and the village of Sommers to the northwest showed the largest increases in total votes cast, while increases in the city center were relatively flat.

Figure 7.5 shows the election results for Joe Biden at the voting precinct level throughout Kenosha as well as the locations of both Jacob Blake's shooting and victims shot by Kyle Rittenhouse. It is clear that Joe Biden's greatest support was in the central areas of the city of Kenosha. His support generally increased in voting districts with larger African American populations in downtown Kenosha and decreased the farther away from the city center.

Comparing Figures 7.4 and 7.5, it is clear increased voter turnout compared to 2016 was generally higher in areas where Trump did better than Biden. Figure 7.6 shows the percent change in vote between 2020 and 2016 for Donald Trump at the voting district level. This figure indicates Trump exceeded his turnout through much of the county, with his largest percentage

Figure 7.4 Change in Total Vote for the 2020 and 2016 Presidential Elections for Kenosha County Voting Districts.

Data source: Wisconsin Election Commission.

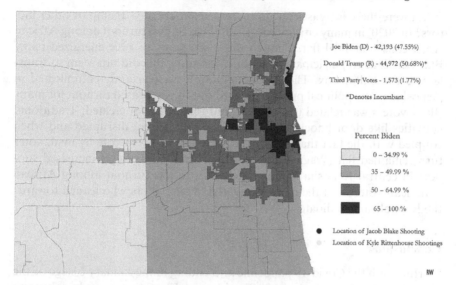

Figure 7.5 Percent Joe Biden Vote for Kenosha County Voting Districts.
Source: Wisconsin Election Commission.

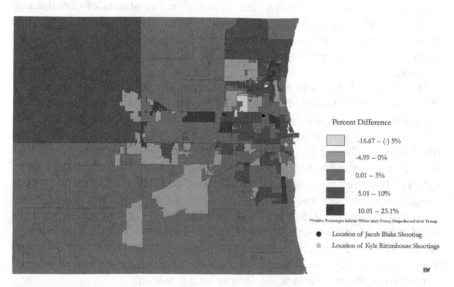

Figure 7.6 Percent Difference in Trump Vote Between 2020 and 2016 for Kenosha
County Voting Districts.

Data source: Wisconsin Election Commission.

increases in many of the voting districts in the city of Kenosha itself.
Specifically, Trump's increased support in downtown Kenosha was located in
areas most heavily impacted by protests that erupted after the shooting of
James Blake.

Yet were these increases due to increased support for Trump or other factors? In 2020, in many cities across the United States, turnout among African Americans remained flat. Though the discussions of race energized many Black citizens to participate in peaceful protests, this did not seem to translate to the ballot box. Though the situation was certainly complicated in Kenosha, many political pundits pointed out the decreased turnout for many Black voters was related to Covid-19 and the hurdles it created. Traditional activities like door-knocking and election rallies were disrupted and when coupled with the fact that Black voters are less likely to vote by mail, voter turn remained flat (Wagtendonk 2020). Therefore, Trump's increased support in central Kenosha is more likely due to flat turnout among African American voters and the more motivated White voters who leaned towards the Republican candidate.

Conclusions

Starting with the Covid-19 pandemic and ending in the January 6 siege of the United States Capitol, the 2020 Presidential Election was truly a historic event. 2020 will also be remembered for Breonna Taylor, George Floyd, Jacob Blake, and the many other victims of police brutality. The protests that started in Louisville in March and violence in the streets of Kenosha in September all became defining moments of the presidential election. The political divide that ensued took sharp partisan tones and would eventually be reflected in the election patterns that emerged after the November 3 election.

It was clear before the events of 2020 that the speeches and actions of President Donald Trump often implied a support for White supremacy. After the initial protests of the George Floyd murder, Trump used his pulpit to side with police and against protestors. Often calling protesters rioters, thugs, and un-American, Trump connected with many White voters by highlighting a fear of violence and looting often portrayed on social media. Although thousands of peaceful protests occurred across the country and the world, violent images of protestors and burning buildings were difficult for many White voters to ignore. For Democrats and Joe Biden, it was easy to embrace the Black Lives Matter movement, but in the days after the initial Floyd protests, Biden would be hampered by slogans calling to "Defund the Police." Though Biden made it very clear he was not in favor of defunding police, it allowed Trump to consistently attack the former vice president.

Before the shootings in Kenosha, Trump's rhetoric was having an impact. Polls showed support for Black Lives Matter protests was waning, especially among White voters. These divides grow deeper on August 23 with the shooting of Jacob Blake and the protests and further shootings by Kyle Rittenhouse on August 25. The entire situation in Kenosha, Wisconsin further highlighted the partisan divides that developed over the topic during the 2020 presidential campaign. The situation forced both candidates to visit Kenosha and bring race and police brutality back into the forefront of the election.

This chapter provides clear evidence that the shooting of Jacob Blake and the aftermath influenced the presidential election results in Kenosha County. Citizens on both sides of the argument expressed frustrations regarding the shootings, the protests, the looting, and the candidates themselves in the days and weeks after Jacob Blake was shot. The visiting by both candidates signaled the clear divide on the topic. While some Kenosha citizens like Michelle Stauder were happy Biden came to visit, others, like Chris Parker, said he never aligned himself with Trump but said that the Democrats have done little about the situation (Milwaukee Journal Sentinel 2020). Though many voters appreciated the visits from Biden and Trump, some voiced concerns both candidates were simply taking advantage of the situation. Jamar Mayfield questioned of Biden, "Is he just trying to get a vote?" (Alter 2020). In the end, it is clear Trump's message of law and order rang loudly among White voters, especially in the suburban areas, while Covid-19 depressed Black voter turnout, especially in the downtown areas of Kenosha.

References

Ainsley, J. 2020. Internal document shows Trump officials were told to make comments sympathetic to Kyle Rittenhouse. *NBC News* (Oct. 1). https://www.nbcnews.com/politics/national-security/internal-document-shows-trump-officials-were-told-make-comments-sympathetic-n1241581

Alter, C. 2020. Joe Biden visits Kenosha. *Time* (Sept. 3). https://time.com/5886059/joe-biden-kenosha-jacob-blake/

Barrett, R. and J. Taschler. 2010. Final chapter in Kenosha's automotive history ends quietly. *Milwaukee Journal Sentinel* (Oct. 22). http://archive.jsonline.com/business/105580433.html/

Barrow, B., W. Weissert, and S. Bauer. 2020. Biden, in Kenosha, says U.S. confronting "original sin". *AP News* (Sept. 3). https://apnews.com/article/election-2020-ap-top-news-race-and-ethnicity-racial-injustice-politics-2373c9517d2781ab11b92e1ee0dd5273

Blake, A. 2020. A slip in support for Black Lives Matter? *Washington Post* (Aug. 29). https://www.washingtonpost.com/politics/2020/08/29/slip-support-black-lives-matter/

Booker, B., and E. Bowman. 2020. Wisconsin deploys National Guard after shooting of Black man sparks protests. *NPR* (Aug. 24). https://www.npr.org/sections/live-updates-protests-for-racial-justice/2020/08/24/905316709/wisconsin-police-shooting-leaves-black-man-in-serious-condition

Carthage College. 2021. Quick facts. https://www.carthage.edu/about/quick-facts/

CNBC. 2020. Trump stokes national divisions in Fourth of July speech. *CNBC* (July 5). https://www.cnbc.com/2020/07/05/trump-stokes-national-divisions-in-fourth-of-july-speech.html

Conway, K. 2020. New Marquette Law School Poll Finds Biden Lead Over Trump Stables at Five Percentage Points (Oct. 28). *Marquette New Center*. https://www.marquette.edu/news-center/2020/new-marquette-law-school-poll-finds-biden-lead-over-trump-stable-at-five-percentage-points.php

Cohen, B. 2020. Milwaukee Bucks' protest of Jacob Blake shooting stops sports world. *Wall Street Journal* (Aug. 26). https://www.wsj.com/articles/milwaukee-bucks-boycott-playoff-game-to-protest-kenosha-police-shooting-11598474937

Eligon, J., S. Mervosh, and R. Oppel, Jr. 2020. Jacob Blake was shackled in hospital bed after police shot him. *New York Times* (Aug. 28). https://www.nytimes.com/2020/08/28/us/jacob-blake-shackles-assault.html

Flores, T. 2020. Damage due to rioting, unrest in Kenosha tops $50 million; 2,000 Guard assisted here. *Kenosha News* (Sept. 9). https://www.kenoshanews.com/news/local/damage-due-to-rioting-unrest-in-kenosha-tops-50-million-2-000-guard-assisted-here/article_26473ec9-c08a-5490-9d09-cc2b840b65f1.html

Franklin, C. 2021. Black Lives Matter protests in Wisconsin. *RPubs*. https://rpubs.com/PollsAndVotes/652966

Great Schools. 2021. https://www.greatschools.org/wisconsin/kenosha/schools/?gradeLevels%5B%5D=h

Guarino, M. 2020. Kyle Rittenhouse released from jail after posting $2 million bail. *Washington Post* (Nov. 20). https://www.washingtonpost.com/nation/2020/11/20/kenosha-shooting-rittenhouse-bail/

Jimenez, O. 2021. Exclusive: Jacob Blake speaks out a year later. "I have not survived until something has changed". *CNN* (Aug. 29). https://www.cnn.com/2021/08/29/us/jacob-blake-exclusive-shooting-a-year-later/index.html

Johnson, A., M. Johnson, and T. Shelbourne. 2020. What we know about the victims of the Kenosha protest shooting that killed two men and injured another. *Milwaukee Journal Sentinel* (Aug. 26). https://www.jsonline.com/story/communities/lake-country/2020/08/26/kenosha-jacob-blake-protest-shooting-what-we-know-victims/5635760002/

Jordan, B. 2020. Kenosha business owner declines President Trump photo-op, former owner replaces him. *TMJ4* (Sept. 1). https://www.tmj4.com/news/local-news/kenosha-business-owner-declines-president-trump-photo-op-former-owner-replaces-him

Lebron, C. 2017. *The Making of Black Lives Matter: A Brief History of an Idea.* Oxford: Oxford University Press.

Liptak, K. 2020. Trump stays silent on Jacob Blake shooting but vows to stop violent protests and slams the NBA. *CNN* (Aug. 27). https://www.cnn.com/2020/08/27/politics/donald-trump-jacob-blake/index.html

Maxouris, C. 2020. The 26-year old man killed in Kenosha shooting tried to protect those around him, his girlfriend says. *CNN* (Aug. 28). https://www.cnn.com/2020/08/28/us/kenosha-wisconsin-protest-shooting-victims/index.html

McCarthy, T. 2020. Trump arrives in Kenosha with Bill Barr as he pushes law-and-order message. *The Guardian* (Sept. 1). https://www.theguardian.com/us-news/2020/sep/01/trump-golf-police-shooting-comparison-fox-news-interview

McLaughlin, E., M. Holcombe, and B. Parks. 2020. Jacob Blake is out of the hospital, but how long he'll be in rehab remains a question. *CNN* (Oct. 7). https://www.cnn.com/2020/10/07/us/jacob-blake-leaves-hospital/index.html

Miller, Z., and J. Lemire. 2020. Trump visits Kenosha, calls violence "domestic terrorism". *AP News* (Sept. 1). https://apnews.com/article/virus-outbreak-election-2020-ap-top-news-politics-shootings-4a58a15c9955bb6312c1fbe42215110d

Milwaukee Journal Sentinel. 2020. Joe Biden in Kenosha: Candidate "felt good about" Wisconsin visit. *Milwaukee Journal Sentinel* (Sept. 3). https://www.jsonline.com/story/news/2020/09/03/joe-biden-kenosha-live-updates-speech-blake-family-visit/5701026002/

Morales, C. 2021. What we know about the shooting of Jacob Blake. *New York Times* (Nov. 16). https://www.nytimes.com/article/jacob-blake-shooting-kenosha.html

Petras, G. and R. Padilla. 2020. A visual timeline of violence in Kenosha after police shooting of Jacob Blake. *USA Today* (Aug. 27). https://www.usatoday.com/in-depth/graphics/2020/08/27/jacob-blake-kenosha-police-shooting-two-killed/3442878001/

Richmond, T. and M. Tarm. 2021. No charges against Wisconsin officer who shot Jacob Blake. *AP News* (Jan. 6). https://apnews.com/article/kyle-rittenhouse-pleads-not-guilty-cf6228f55a4f2a5fdd66978e5523a912

Rose, A., S. Sidner, and O. Jimenez. 2020. Multiple buildings in Kenosha were set on fire tonight. *CNN* (Aug. 25). https://www.cnn.com/us/live-news/kenosha-wisconsin-police-shooting/index.html

Smith, D. 2020. Sex assault charge against Jacob Blake dropped; he pleads guilty to disorderly conduct. *Kenosha News* (Nov. 6). https://www.kenoshanews.com/news/local/sex-assault-charge-against-jacob-blake-dropped-he-pleads-guilty-to-disorderly-conduct/article_57a11cb6-1516-521c-9d42-73366c7eb4ab.html

Timberg, C. 2020. Zuckerberg acknowledges Facebook erred by not removing a post that urged armed action in Kenosha. *Washington Post* (Aug. 28). https://www.washingtonpost.com/technology/2020/08/28/facebook-kenosha-militia-page/

Treisman, R. 2020. Kenosha shooting suspect faces homicide charges in protestors' deaths. *NPR* (Aug. 27). https://www.npr.org/sections/live-updates-protests-for-racial-justice/2020/08/27/906901940/kenosha-shooting-suspect-charged-with-six-criminal-counts-including-homicide

Treisman, R. 2021. Jacob Blake files excessive force lawsuit against Kenosha police officer who shot him. *NPR* (March 26). https://www.npr.org/2021/03/26/981590041/jacob-blake-files-excessive-force-lawsuit-against-kenosha-police-officer-who-sho

United States Census. 2019. American Community Survey. https://www.census.gov/quickfacts/fact/table/kenoshacitywisconsin/PST045219

University of Wisconsin-Parkside. 2021. Emerging Hispanic Serving Institution. https://www.uwp.edu/explore/aboutuwp/hsi/hsi.cfm

Vielmetti, B. 2020. Kenosha car dealer denies he asked gunmen to protect his business. *Milwaukee Journal Sentinel* (Sept. 3). https://www.jsonline.com/story/news/crime/2020/09/03/kenosha-car-dealer-denies-he-asked-gunmen-protect-his-business/5705974002/

Wagtendonk, A. 2020. Milwaukee's black turnout down in 2020. *Urban Milwaukee* (Nov. 13). https://urbanmilwaukee.com/2020/11/13/milwaukees-black-turnout-down-in-2020/

Wong, W. 2021. Kyle Rittenhouse, out on bail, flashed white power signs at a bar, prosecutors say. *NBCNews* (Jan. 14). https://www.nbcnews.com/news/us-news/kyle-rittenhouse-out-bail-flashed-white-power-signs-bar-prosecutors-n1254250

Willis, H., M. Xiao, C. Triebert, C. Koettl, S. Cooper, D. Botti, J. Ismay, and A. Tiefenthaler. 2020. Tracking the suspect in the fatal Kenosha shootings. *New York Times* (Aug. 27). https://www.nytimes.com/2020/08/27/us/kyle-rittenhouse-kenosha-shooting-video.html%20

8 Diseased Democracy
Geographies of the Covid-19 Pandemic and its Impact on the 2020 Presidential Election

Shaun J. Johnson

The Covid-19 pandemic has wreaked havoc on every facet of the American political system since the virus traveled to the United States in early 2020. Its emergence from the Wuhan province of China was regrettable, although, as many epidemiologists have argued, largely predictable (Berezow et al. 2018; Doucleff 2021). The occurrence of this new viral plague during a tightly contested election year exposed just how vulnerable routine democratic processes could be to a severe outbreak of disease. Indeed, the Covid-19 pandemic overrode all other issues during the 2020 presidential election, turning attention away from policy and toward disaster response.

During the quadrennial presidential election cycle the response to the Covid-19 pandemic took front and center. The incumbent president, Donald Trump, had argued throughout the course of the pandemic against the restrictions recommended by the Centers for Disease Control (CDC), often even contradicting administrators after they had spoken in official press briefings (Siemasko 2020). Trump also frequently downplayed the severity of the virus, saying multiple times throughout the course of 2020 that the pandemic would "fade away" or that he was "unconcerned" about the virus (Al Jazeera 2020). This rhetoric from the president appealed to his base, and many of his voters followed his lead on Covid-19 precautions, with the Pew Research Center finding in June 2020 that 61% of Republican voters thought the worst of the pandemic was behind us. While appealing mostly to his fervent base, the anti-health restriction and anti-mask rhetoric from Trump was generally his only case for re-election: a return to business as usual and a general agreement to ignore the issues of the pandemic. While it might seem crass, a "balanced ship" message, even if not delivered upon by the president, has won re-election before, and, as will be discussed later, Trump's position of being the incumbent provided many advantages in this kind of campaign (Bush 2019; Mayhew 2008).

On the opposite side of the political aisle, the Democratic Party was looking to mount a successful challenge to the incumbent president. Candidates from across the political spectrum entered into the battle of the presidential primaries, with the main ideological fight being between more progressive candidates such as Senators Bernie Sanders (I-VT) and Elizabeth Warren (D-MA) and moderate Democrats such as former Vice President Joe Biden

DOI: 10.4324/9781003260837-8

and Senator Amy Klobuchar. Despite early difficulties for the Biden campaign and a strong challenge from the seemingly invigorated progressive left wing of the party, the former vice president managed to rally establishment Democrats and by April 8 was their presumptive nominee for president (Ember 2020; Korecki and Siders 2020). The Biden campaign had run on its candidate's perceived electability to the highest office, and this message resonated with many primary voters tired after four years of the Trump presidency (Skelley et al. 2020).

This chapter focuses on how the Covid-19 pandemic impacted the presidential race, utilizing a review of other sources in this area and a geographical analysis of the voting patterns. Several factors will be considered, such as the impact of mail-in voting and allegations of voter fraud, as well as some general overview of past factors that have impacted presidential races. While geographers have, of course, analyzed presidential elections before (see Warf 2006; Watrel et al. 2018), the Covid-19 pandemic provides a unique case study to analyze the impact of major health events on these types of elections.

Electoral Factors

The science of understanding elections, specifically American elections, has become a massive endeavor, spawning academic fields such as electoral geography and political science. Electoral geographers have pointed to a wide variety of factors that impact elections in the past, such as income, voter location, the state of the economy, religiosity, gerrymandering, and racial difference (Watrel et al. 2018). When looking to understand the impact of the Covid-19 pandemic on this election, it is important to look into and rule out some other issues that may have swayed the American voters.

One major factor affecting this presidential bout were the built-in advantages of the incumbent. Despite some evidence of its declining effect on Congressional elections (Jacobson 2015), the incumbent advantage has been shown to support presidential reelection many times throughout US history (Mayhew 2008). Going into this presidential contest, Donald Trump enjoyed the advantage of having a huge media presence that allowed his campaign to shape the course of debate, and being in the political spotlight for four years provided the American people plenty of exposure to his candidacy. If Trump were to lose the 2020 presidential election, he would be the first president since President George Bush Sr. to not win re-election (in 1992), and so the stakes for his campaign were high (Alpert 2022).

Despite the great amount of press attention the Trump campaign received, the press was often negative or accusatory. Trump had long battled with the media, lobbing accusations against popular TV programs and accusing multiple news sources of being broadly "fake news" (Kurtz 2018). This negative press and the incumbent president's wild conspiracy theorizing struck a sharp contrast with the message being put out by the Biden campaign, one of calm, normalcy, and a return to common ground. Over the course of his campaign, the former vice president emphasized the need to both get the Covid-19 pandemic under control, but also expressed a desire to restore a political landscape of bipartisan cooperation (Detrow and Khalid 2020; Bolsover 2020; Hart 2021). The Biden campaign stressed from the beginning that it was running to save the "soul of the nation" from a president who had gone off the rails in the minds of many voters (Detrow and Khalid 2020). This argument appealed to a portion of American voters who were tired of the constant political drama, and some political analysts were arguing in the run-up to the election that this alone would be enough to topple the Trump presidency.

Another factor affecting the re-electability of Donald Trump was the state of the American economy. The economy has long been a factor that many political scientists consider to be of utmost importance to incumbent electability (Fair 1996, 2009; Formisano 2010). Going into the 2020 presidential race, the economy was ailing from months of shutdowns due to Covid-19. The rate of unemployment, a longtime measure of the health of the American economy, had spiked to its highest point in modern history of 14.8% in April 2020 amid waves of business closures and Covid-19 restrictions. Though unemployment had dropped back to 6.9% by October 2020, this was still slightly above the US average of around 5.7% (Bureau of Labor Statistics 2021), and economic factors have been shown to greatly influence presidential election results in the past, and even influence party identification at the ballot box (De Neve 2014; Guntermann, Lenz, and Myers 2021). With this in mind, Republican-led state governments across the country began to push for an economic reopening in early summer 2020, urged on by the Trump re-election campaign (Roberts 2020). Trump campaign messaging reinforced the idea that the economy *had* been great and would still be if not for the "China virus" and urged voters to support an end to Covid-19 restrictions. This urge to reopen the economy came despite numerous reports that doing so too soon might lead to thousands more casualties, a fact brushed aside and downplayed by Trump administration officials (Superville and Lemire 2021).

This breezy attitude toward the loss of life may have come at a cost of votes for the incumbent president. Casualties that have occurred under the watch of past presidents have had a negative impact on voters, with evidence suggesting that counties with higher rates of wartime casualties in the Iraq and Afghanistan conflicts had higher rates of voting against the incumbent administration (Kriner and Shen 2007; Gartner 2008). Though the Trump campaign may have believed downplaying Covid-19 deaths and pressuring the reopening of the economy would work in its favor, evidence from July 2020 showed his likely voters' approval rating dipping in heavily impacted counties (Warshaw et al. 2020). These results were borne out after the election as well, with Baccini et al. (2021) noting that "A simple counterfactual analysis suggests that Trump would likely have won re-election if COVID-19 cases had been 5 percent lower." Though this analysis makes sense in terms of statistical analysis, on the maps it shows a stranger picture.

Another factor to consider in this election is that the pandemic resulted in an unprecedented increase in mail-in voting. The practice has its origin in the civil war, when soldiers were given the option to vote by mail for the first time while away from their home states (Moreton 1985). This beginning informed the general structure of mail-in voting for the next hundred or so years of American democracy, with the practice being predominately used by those unable to make it physically to the ballot box due either to illness or to military service. In this election, with concerns over Covid-19 looming over the election process, voters were more likely to vote by mail than ever before. Early on in the election cycle, however, it became clear that the Trump campaign was determined to demonize mail-in voting, with targeted language in speeches and ad campaigns calling it fraudulent and claiming the process was not secure (Saul and Epstein 2020; Parks 2020). This demonization of the process resulted in a large difference between mail-in votes cast for Biden and Trump, with the differences between the two becoming obvious as the election results rolled in following the election. For example, the Pew Research Center found that Biden voters were far more likely to vote by mail than Trump voters, 58% for Biden vs 32% for Trump (Pew Research Center 2020). Further, the Pew report also indicated that many voters returned their mail-in votes early, with about 76% returned before election week, reflecting a widespread desire to get votes in well before potential cut-off dates.

Despite the obvious advantages of mail-in voting during the Covid-19 pandemic, many Trump voters still went to the polls on election day, resulting in an early vote bump for the Trump–Pence ticket. This early advantage led the former president to declare his victory within just a few hours of the polls closing, despite millions of ballots being left to count (Itkowitz et al. 2020). The initial in-person vote followed by the late surge in mail-in votes also set the ground for former President Trump to decry the results of the election again when tallying was complete, claiming the results were faulty and that the election had been stolen. This demonization of mail-in voting by the former president is an abrupt departure from past voting patterns for Republican candidates. While there is conflicting evidence as to whether voting by mail actually increases turnout, in this election the partisan results of the practice were nearly negligible, and voting by mail did not appear to increase votes cast for any particular candidate (Barber and Holbein 2020; Fitzgerald 2005). Despite far more Democratic voters casting their ballots by mail, this was compensated for by a lack of Democrats at the ballot box on election day, and so the effects of this practice can be discounted.

Geographical Analysis

To tease out the impacts of Covid-19 on the election, it is worthwhile to look at how Covid-19 affected the voting in each individual county on election day. A county-level look at the issue allows for easy comparison to Covid-19 case data without sacrificing the detail that could be obscured by

County Level Voting for Each Candidate

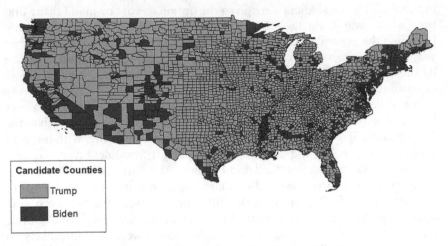

Figure 8.1 County-level Election Results, 2020 Presidential Election.

Source: Voting data from MIT Election lab 2020; figure by author.

a state-to-state comparison. The data for this project were taken directly from the Centers for Disease Control and Prevention and the Massachusetts Institute of Technology Election Data and Science Lab, two reliable sources for election study.

Figure 8.1 shows voting outcomes by American counties in the 2020 presidential election. Each county is colored along partisan lines with the results of their voting tally in the election, with the traditional Republican red signifying Trump counties and the Democrat blue signifying Biden counties. As can be easily seen, a majority of counties in the United States voted for Trump, though this represents more an abundance of land rather than an abundance of voters. The traditional red and blue America map has been analyzed numerous times in electoral geography (see Mellow and Trubowitz 2005; Morrill et al. 2007), though it is worth noting here that 25 former Trump voting counties were picked up by Biden during the election, mainly in swing states (Ballotpedia 2021). This pickup was actually less than what could be expected for the presidential candidate, with nearly 181 counties that voted for Obama in 2012 voting for Trump in 2020.

As has been pointed out, the American public has traditionally held presidents to account for casualties that occur during their presidency (Gartner 2008). This would seem to indicate that those counties with higher casualties would vote at a lower rate for the incumbent president. To most accurately narrow down the impacts of the pandemic on vote change the Covid-19 data were limited to cases which had occurred up to election day on November 3, 2020. The totals were then taken and turned into case rates per 100,000

Cases by County, November 3rd, 2020

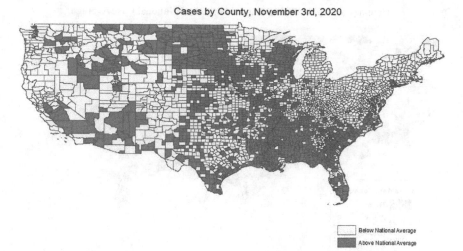

Figure 8.2 Counties Above and Below the National Average Covid Caseload, November 2020.

Source: Data from the CDC, 2020; figure by author.

people, a standard CDC measure, to account for the issue of disproportionate population across counties. The average caseload per county on election day 2020 was 2,949.96 cases per 100,000 people, simplified to 2,950 cases per 100,000 for the purposes of easier mapping.

Figure 8.2 details those counties that have a caseload above the national average and it shows an interesting pattern. As can be seen, counties in the South and Midwestern regions of the country were hardest hit by Covid-19. Notably major metropolitan areas like New York and Chicago were above the national average, but other urban areas, such as Los Angeles, Seattle, and San Francisco, were not. There was also a mixed bag of electorally vital swing states in the mix on election day, with Pennsylvania, Michigan, and Nevada ranking well below the average but Wisconsin, Florida, and Georgia inundated with cases.

To examine whether the pandemic had an impact on voting, it is useful to combine both of these metrics. Figure 8.3 shows all counties that voted for either candidate, with an emphasis on those counties which were above the national average caseload. As can be seen from the map, counties that were above the national average caseload on November 3, 2020 were more likely to vote for Trump than for Biden. Those counties above the average that did vote for Biden were predominately in Southern and swing states, and if the state to which they belonged voted as a whole for the Democratic candidate, they were vital in that tally (Apache County Arizona, for example).

The combination of these factors presents an interesting picture when we consider again our outside sources. Though Biden picked up five swing states

Counties Voting for Either Candidate, Above National Case Average

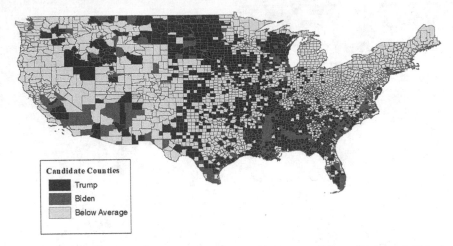

Figure 8.3 Counties voting for either candidate with a higher-than-average Covid-19 rate.

Source: Data from the CDC and the MIT Election Science Lab; figure by author.

to win the election, his gains were kept almost exclusively in urban and suburban areas. Though Covid-19 casualties may have caused voters to turn away from former-President Trump's policies, these same motivations seem not to have resonated with the rural electorate.

Analysis and Conclusion

With the election over and the Biden presidency entering its second year, it is a good time to take a step back and analyze some of the factors that put him in office. While the former president put up a good fight for the office, there were some factors that blocked his return to the resolute desk. Using the figures presented above, there are several hypotheses about whether or not the Covid-19 pandemic swayed voters away from his candidacy.

The first theory is that the economy was a larger factor playing into the minds of voters than Covid-19. Republican administrations have traditionally been seen by voters as better for the economy, though there is abundant evidence that points to this assertion as being untrue (Alesina and Rosenthal 1995; Amadeo 2021; Sheffield 2019). While polling done well before the election had shown that containing the Covid-19 pandemic and getting the vaccine out were top priorities, as the election drew nearer the issue began to be sidelined by economic recovery. In August of 2020, with only a few months until the vote, 62% of voters indicated that the Covid-19 outbreak would be a top factor in their voting decision (Pew Research Center 2020). By the time the exit polls were rolling out on election day that number had dropped to just around 20% of voters, with most of those being likely-Biden voters already (*Washington* Post 2020). This could explain why although Covid-19 was an issue on many voters' minds during the election, containing the pandemic did not sway quite as large a wave of voters away from Trump as might be expected.

A second hypothesis is that those voters who were more exposed to casualties were more likely to take the Covid-19 pandemic seriously. While obvious on its face, this theory could explain why, though many counties had above-average rates of Covid-19 transmission, a general lack of population centers made this less of an issue to these voters. For example, as can be seen in Figure 8.3, rural counties in the Dakotas were well above the national average caseload; however, they voted overwhelmingly for the incumbent president. This trend is reversed in suburban areas near larger cities like New York and Chicago, where caseload is high and voters voted at much greater rates for Biden in 2020 than Hillary Clinton in 2016. As has been said many times by political geographers and political scientists alike, whichever candidate carries the suburbs wins the election, and there is a clear relation to that theory in this election.

Overall, this election showed that major health events can easily influence the outcome of important elections. When coming up against an incumbent candidate, whatever their personality and whatever the position, the challenger faces significant difficulties in overcoming that inherent advantage. For President Biden, the Covid-19 pandemic was enough to lift him into office over his opponent. This analysis has shown that while Covid-19 may not have shaken the most loyal of Trump voters in the rural areas of the country, it was enough to turn out the suburban voters who often decide American elections.

References

Al Jazeera. 2020. Timeline: President Trump's comments on the coronavirus. (Oct. 3). https://www.aljazeera.com/news/2020/10/3/president-trumps-comments-on-the-coronavirus-timeline.

Alesina, A. and H. Rosenthal. 1995. *Partisan Politics, Divided Government, and the Economy*. Cambridge: Cambridge University Press.

Alpert, Gabe. 2022. "Presidents Who Didn't Win a Second Term." *Investopedia*. 2022. https://www.investopedia.com/financial-edge/0812/5-presidents-who-couldnt-secure-a-second-term.aspx.

Amadeo, K. 2021. Democrat or Republican: Which political party has grown the economy more? *The Balance* (Oct. 13). https://www.thebalance.com/democrats-vs-republicans-which-is-better-for-the-economy-4771839.

Ballotpedia. 2021. Election results, 2020: Pivot counties in the 2020 presidential election. *Ballotpedia*. https://ballotpedia.org/Election_results,_2020:_Pivot_Counties_in_the_2020_presidential_election.

Barber, M. and J. Holbein. 2020. The participatory and partisan impacts of mandatory vote-by-mail. *Science Advances* 6(35):eabc7685. https://www.science.org/doi/10.1126/sciadv.abc7685.

Baccini, L., A. Brodeur, and S. Weymouth. 2021. The COVID-19 Pandemic and the 2020 US Presidential Election. *Journal of Population Economics* 34(2): 739–767. https://doi.org/10.1007/s00148-020-00820-3.

Berezow, A., H. Campbell, J. LeMieux, and S. Schow. 2018. The next plague and how science will stop it. American Council on Science and Health. https://www.acsh.org/sites/default/files/Next%20Plague.pdf.

Bolsover, G. 2020. COVID's impact on the US 2020 election: Insights from social media discourse in the early campaign period. *SSRN Electronic Journal.* https://doi.org/10.2139/ssrn.3714755.

Bureau of Labor Statistics. 2021. "Civilian Unemployment Rate." 2021. https://www.bls.gov/charts/employment-situation/civilian-unemployment-rate.htm#.

Bush, D. 2019. Where President Trump stands on the issues in 2020. *PBS NewsHour* (June 19). https://www.pbs.org/newshour/politics/where-president-trump-stands-on-the-issues-in-2020.

De Neve, J. 2014. Ideological Change and the Economics of Voting Behavior in the US, 1920–2008. *Electoral Studies* 34 (June): 27–38. https://doi.org/10.1016/j.electstud.2013.10.003.

Detrow, S., and A. Khalid. 2020. Can Joe Biden's Campaign Message Carry Him Over The Finish Line? NPR, November 2, 2020, sec. *Politics.* https://www.npr.org/2020/11/02/930234714/can-joe-bidens-campaign-message-carry-him-over-the-finish-line.

Doucleff, M. 2021. Next pandemic: Scientists fear another coronavirus could jump from animals to humans. *NPR.Org* (March 19). https://www.npr.org/sections/goatsandsoda/2021/03/19/979314118/next-pandemic-scientists-fear-another-coronavirus-could-jump-from-animals-to-hum.

Ember, S. 2020. Bernie Sanders drops out of 2020 Democratic race for president. *New York Times* (April 8). https://www.nytimes.com/2020/04/08/us/politics/bernie-sanders-drops-out.html.

Fair, Ray C. 1996. Econometrics and presidential elections. *Journal of Economic Perspectives* 10(3): 89–102. https://doi.org/10.1257/jep.10.3.89.

Fair, Ray C. 2009. Presidential and Congressional vote-share equations. *American Journal of Political Science* 53(1): 55–72.

Fitzgerald, M. 2005. Greater convenience but not greater turnout: The impact of alternative voting methods on electoral participation in the United States. *American Politics Research* 33(6):842–867.

Formisano, Ron. 2010. Populist currents in the 2008 presidential campaign. *Journal of Policy History* 22(2): 237–255. https://doi.org/10.1017/S0898030610000059.

Gartner, S. 2008. Ties to the dead: Connections to Iraq war and 9/11 casualties and disapproval of the president. *American Sociological Review* 73(4) 690–695.

Guntermann, E., Gabriel S. Lenz, and Jeffrey R. Myers. 2021. The impact of the economy on presidential elections throughout US history. *Political Behavior* 43(2): 837–857. https://doi.org/10.1007/s11109-021-09677-y.

Hart, Roderick P. 2021. Why Trump lost and how? A rhetorical explanation. *American Behavioral Scientist* March, 0002764221996760. https://doi.org/10.1177/0002764221996760.

Itkowitz, Colby, Felicia Sonmex, John Wagner, Derek Hawkins, Paulina Firozi, Meryl Kornfield, David Weigel, and Amber Phillips. 2020. Trumep falsely asserts election fraud, claims a victory. *Washington Post* (Nov. 3) https://www.washingtonpost.com/elections/2020/11/03/trump-biden-election-live-updates/.

Jacobson, G. 2015. It's nothing personal: The decline of the incumbency advantage in US House elections. *Journal of Politics* 77(3):861–873.

Korecki, N. and D. Siders. 2020. Sanders sends democratic establishment into panic mode. *Politico* (Feb. 23). https://www.politico.com/news/2020/02/23/sanders-democratic-establishment-panic-mode-117065.

Kriner, Douglas L., and Francis X. Shen. 2007. Iraq casualties and the 2006 Senate elections. *Legislative Studies Quarterly* 32 (4): 507–530.

Kurtz, Howard. 2018. *Media Madness: Donald Trump, the Press, and the War over the Truth.* Simon and Schuster.

Mayhew, D. 2008. Incumbency advantage in U.S. presidential elections: The historical record on JSTOR. https://www-jstor-org.www2.lib.ku.edu/stable/20203009?pq-origsite=360link&seq=1#metadata_info_tab_contents.

Mellow, N. and P. Trubowitz. 2005. Red versus blue: American electoral geography and Congressional bipartisanship, 1898–2002. *Political Geography* 24(6):659–677.

Moreton, E. 1985. Voting by mail note. *Southern California Law Review* 58(5):1261–1282.

Morrill, R., L. Knopp, and M. Brown. 2007. Anomalies in red and blue: Exceptionalism in American electoral geography. *Political Geography* 26(5):525–553.

Parks, M. 2020. Ignoring FBI And Fellow Republicans, Trump Continues Assault On Mail-In Voting. NPR, August 28, 2020, sec. *Fact Check*. https://www.npr.org/2020/08/28/906676695/ignoring-fbi-and-fellow-republicans-trump-continues-assault-on-mail-in-voting.

Pew Research Center. 2020. Important Issues in the 2020 Election. *Pew Research Center—U.S. Politics & Policy* (blog). (Aug. 13). https://www.pewresearch.org/politics/2020/08/13/important-issues-in-the-2020-election/.

Roberts, N. 2020. As COVID-19 Spreads, Manhattan's Chinatown Contemplates a Bleak Future. Marketplace (blog). March 16, 2020. https://www.marketplace.org/2020/03/16/as-covid-19-spreads-manhattans-chinatown-contemplates-a-bleak-future/.

Saul, S., and Reid J. Epstein. 2020. Trump is pushing a false argument on vote-by-mail fraud. Here are the facts. *New York Times* (Sept. 28) https://www.nytimes.com/article/mail-in-voting-explained.html.

Sheffield, Matthew. 2019. Poll: GOP enjoys 12-point edge over Dems on economy. *TheHill* (April 11). https://thehill.com/hilltv/what-americas-thinking/438482-gop-enjoys-12-point-advantage-on-handling-economy.

Siemasko, C. 2020. Dr. Fauci contradicts Trump's false claim that Covid-19 is as deadly as flu. *NBC News* (Oct. 6). https://www.nbcnews.com/news/us-news/dr-fauci-contradicts-trump-s-false-claim-covid-19-deadly-n1242340.

Skelley, G., L. Bronner, and N. Rakich. 2020. What we know about the voters who swung Super Tuesday for Biden. *FiveThirtyEight* (blog) (March 6). https://fivethirtyeight.com/features/what-we-know-about-the-voters-who-swung-super-tuesday-for-biden/.

Superville, D., and J. Lemire. 2021. Trump Pushes Economy Reopening, Says Virus Could Kill 100K. *APNEWS*. May 1, 2021. https://apnews.com/article/economy-virus-outbreak-donald-trump-politics-business-f3bd02ca4cabd99fc0a5dc4ed332f518.

Warf, B. 2006. Voting technologies and residual ballots in the 2000 and 2004 presidential elections. *Political Geography* 25(5):530–556.

Warshaw, C., L. Vavreck, and R. Baxter-King. 2020. The effect of local COVID-19 fatalities on Americans' political preferences. *Science Advances*. http://chriswarshaw.com/papers/covid_elections.pdf.

Washington Post. 2020. Exit poll results and analysis for the 2020 presidential election. (Dec. 14). https://www.washingtonpost.com/elections/interactive/2020/exit-polls/presidential-election-exit-polls/.

Watrel, R., R. Weichelt, F. Davidson, J. Heppen, E. Fouberg, J. Archer, R. Morrill, F. Shelley, and K. Martis. 2018. *Atlas of the 2016 Elections*. Lanham, MD: Rowman & Littlefield.

9 "The Apartment above a Meth Lab"?

Participation in and Impacts of the 2020 U.S. Election in Canada

Jamey Essex

A widely shared meme that circulated on social media ahead of the 2020 U.S. general election proposed that Canadians must "feel like they live in the apartment above a meth lab." The joke, of course, is that the U.S., like a methamphetamine lab, makes for a bad neighbor: full of toxic chemicals and people, and unpredictably explosive and dangerous.[1] For Americans lamenting Donald Trump's presidency and re-election campaign and the state of political discourse in their country more generally, the joke worked because it affirmed that for foreign observers, something was deeply wrong with the U.S. and needed fixing (no word, however, on whether Mexico felt it was living *below* a meth lab). For Canadians, the humor came in the tension between the internal and external positions the meme highlighted. On the one hand, there is the comfortable reassertion that Canada is different from the U.S.: stable, predictable, and always a good neighbor even if it found itself in a bad neighborhood. On the other hand, meth labs do sometimes explode, and Canadian observers knew the election outcome could have serious and perhaps negative consequences for the U.S.–Canada relationship and for Canadian politics.

The U.S. election remained front and center in Canadian news media for months prior to and in the tumultuous weeks after November 3, 2020, as Donald Trump contested his loss to Joe Biden. This is not surprising, as the U.S. is Canada's largest and most important trading partner and the two countries share a border thousands of miles long, cooperate on numerous political and economic fronts, and share multiple cultural affinities, including the growth of far right and White nationalist organizations, as I discuss below. Hundreds of thousands of Canadians also have family or own property in the U.S., and more than 620,000 potential American voters live in Canada, though their turnout for American elections is historically very low, in the single digits as a percentage of eligible voters (FVAP 2016, 2018, 2020; Porter Robbins 2020). While this is within the normal range for turnout among overseas American voters, the stakes of the 2020 election were unusually high for Americans in Canada because the two countries closed their land border to non-essential travel as part of pandemic-related public health and security measures starting in March 2020. Beyond the pressures of controlling coronavirus, the election highlighted several important trends in the

DOI: 10.4324/9781003260837-9

U.S.–Canada relationship, for American voters abroad, and for the influence (and limits) of American political behavior and rhetoric on its northern neighbor.

I examine three such trends in this chapter: first, how expatriate American voters in Canada participated in the 2020 election cycle; second, how border restrictions related to the coronavirus pandemic shaped economic and social connections in key cross-border regions, especially Windsor-Detroit, and made the border an election issue; and finally, how the tactics and tenor of the Trump campaign have influenced political messaging in Canada, especially in terms of nationalist and populist rhetoric and opening space for far-right groups and messages. Before getting to these three themes, I first examine the broader context of extraterritorial voting and geographic considerations of voting by mail, which dominated the administration of the 2020 election domestically but have long been the norm for extraterritorial American voters. I also should note that I am an American voter and dual American-Canadian citizen residing in Canada who voted by absentee ballot in the 2020 election, and my own experience of voting by mail in numerous elections informs part of the discussion that follows.

Context: Extraterritorial Voting and Voting by Mail

Understanding how the 2020 U.S. election mattered for Americans voting from Canada, and for Canadian politics and the U.S.–Canada relationship, requires context in relation to two themes in the extant literature on voting: first, extraterritorial voting and how the U.S. system compares to those in other countries; and second, the rapid evolution of voting by mail in U.S. domestic elections. Extraterritorial voting has expanded considerably around the world in the last three decades, with 70 percent of states allowing some form of extraterritorial voting by citizens living abroad in 2017, up from 15 percent in 1980 (Burgess and Tyburski 2020, p. 2). This phenomenon reflects not just global migration but states' and political parties' efforts to cultivate blocs of expatriate voters who are often more educated, wealthier, and mobile than the national average. Some states have even allowed for direct representation of expatriate voters in national legislatures, while others, including the U.S., count their votes as part of existing domestic jurisdictions. Global trends suggest that although liberalization of voting rights has increased rapidly, many expatriate voters do not participate in domestic elections in their countries of origin (Burgess and Tyburski 2020). Individual costs of voting from abroad in terms of time and effort partly explain this low turnout, while institutional, political, and social contexts matter greatly for both turnout of expatriate voters in specific electoral contests and for their sustained connection to and participation in national and transnational political communities (Ciornei and Østergard-Nielsen 2020; Collyer 2014; Hartmann 2015; Jaca and Torneo 2020; Leal et al. 2012).

Domestic political parties have also sought to mobilize emigrant voters, though this is highly uneven across national contexts, dependent on parties' electoral incentives and organizational structures, and the political inclinations of expatriate voters (Rashkova and Van Der Staak 2020). Kernalegenn and Pellen (2020, p. 111), for example, demonstrate how Emmanuel Macron's En Marche! (EM) quickly cultivated and mobilized a significant number of expatriate voters, contributing to Macron's 2017 presidential election victory, and to a steady, if incomplete "normalization over time of the political space of French nationals abroad." The same could be said across other national and diasporic contexts where expanded voting rights, increased party mobilization, and local organization of expatriate voters has occurred. Yet this complex and emergent transnational and extraterritorial political space, stretching across multiple jurisdictions, and political and social commitments, presents inherently geographic challenges and contradictions for expatriate and emigrant voters. Kull (2008, p. 460) states that while "democratic institutions generally presume that some combination of *citizenship* and *location* provides sufficient proxies for *interest* and *identity*," interests and identities are not geographically or spatially bound in the same ways as citizenship and location. Global migration and the extension of voting rights to émigrés and expatriates mean that the obligations and benefits of citizenship and location are not coterminous within the bounds of the nation-state and sub-national administrative units. Overseas voters face costs to voting with no guarantee that the benefits of electoral participation and representation will follow, while Tager (2006, p. 36) cautions that states may also seek to restrict or channel mobilization of such voters to prevent "questions … about the legitimacy of an election decided by expatriate voters."

The legitimacy of mailed ballots became a key issue in the 2020 U.S. election, as 46 percent of American voters cast their ballots by mail, with another 26 percent voting early in-person and only 28 percent voting in-person on Election Day itself (Stewart 2020). With so many domestic voters choosing to vote by mail amid the pandemic, Donald Trump and many other Republican officials and lawmakers advanced false claims about the provenance and legality of these votes. Stewart (Stewart 2020, p. 4) notes that while in 2008, 2012, and 2016, use of mail voting was roughly even among Republican and Democratic voters, in 2020, almost 60 percent of Democratic voters cast ballots by mail compared to only 30 percent of Republican voters, no surprise given the president's constant drumbeat about mail-in ballots being used to "steal" the election. Donald Trump made his first claims about fraudulent mail ballots and procedures in April 2020, months before the general election. This built, however, from longer-term trends in which fears of rampant voter fraud, often but not always from Republican voters and officials and intensifying since the disputed 2000 presidential election, have brought the legitimacy of American elections into question (Minnite 2011). While claims of widespread election fraud are unfounded, and confidence in voting by mail among those choosing this method was high in 2020 and in line with previous election years (Stewart 2020), this points to the second thread in the literature, which examines the expansion of mail balloting in the U.S. since the 1990s.

Two related points from this literature are instructive for understanding how the 2020 election impacted American voters *outside* the U.S., including those in Canada. First, the general expectation with expanded mail balloting is that it will increase voter turnout, though studies of mandatory mail voting in California (Bergman and Yates 2011; Dubin and Kalsow 1996; Elul et al. 2017) and in Oregon (Karp and Banducci 2000) have shown this is not necessarily the case and that numerous other factors may affect turnout in mail-only elections. Second, the geographic context of mail balloting strongly shapes how absentee voters engage with the electoral process. Gimpel et al. (2006), for example, find that time-constrained suburban voters are most likely to use "convenience voting" methods when available, such as early in-person, mail-in, and no-excuse absentee voting. Yet mail-in balloting for extraterritorial voters can also be time-consuming because of the lag in requesting, receiving, and submitting a completed ballot, as well as the time spent in securing international postage. In locations where voters must rely on second countries' postal services and cross-border processing of mail, the convenience of mail-in balloting is limited by the extent of national territory and associated service handoffs. Baringer et al. (2020, p. 292) also note that voting by mail "is uniquely vulnerable to administrative discretion" on issues such as signature matching, resulting in potentially high rates of ballot rejection. While many domestic voters have the opportunity to "cure" their ballots if problems arise, extraterritorial voters typically cannot return to a local election office to fix ballot problems found after submission.

The coronavirus pandemic's impacts on the U.S.–Canada border regime and active attempts by Trump and other Republican leaders to cast doubt on mailed ballots made the 2020 election particularly difficult for American voters in Canada, exacerbating the costs and challenges of extraterritorial voting. For these voters, this meant starting the voting process early to ensure votes were submitted and counted on time, and a marked increase in registration and "get-out-the-vote" efforts by partisan and non-partisan groups. Yet as American voters in Canada became the target of vigorous voter mobilization and outreach efforts, the candidates said little about specific issues that might matter for Americans living in Canada or overseas more broadly, even after the Covid-related closure of the U.S.–Canada border.

American Voters in Canada

As stated, approximately 620,000 potential U.S. voters reside in Canada, the largest number of extraterritorial American voters in any one foreign country. The population of eligible U.S. voters in Canada and around the world is too socially and spatially diverse, and turnout too low and unpredictable, for confident generalization. In addition, their votes are not counted as a bloc (i.e., there is no Electoral College vote for "Americans abroad") but feed into vote totals in state and local jurisdictions where voters are registered in the U.S. These numbers are not usually or easily separable, so it is difficult to determine how many American voters living in Canada and registered in a particular state or county voted, or for whom. With the possible exception of active-duty members of the U.S. military, Americans abroad do not typically stand out as a targetable subset of voters for political campaigns, and they lack a coherent set of interests as a constituency, save for federal tax filing obligations and absentee voting regulations. Yet the heated rhetoric, potentially thin margins, and relatively surprising outcome of the 2016 election spurred widespread mobilization efforts targeting extraterritorial American voters in 2020. In addition, most states in the U.S. expanded mail balloting to deal with the coronavirus pandemic, increasing and highlighting strains on the postal system and local election administration, difficulties that overseas voters already faced.

Before examining American voters in Canada and U.S.-based organizations mobilizing them, I offer a personal and individual reflection on extraterritorial voting in 2020. As disclosed above, I am myself an American voter in Canada, registered since 1995 in Kentucky, where I was born and grew up. I have voted absentee in Kentucky after moving from the state in 1999, and since 2005 as an American citizen residing in Canada and then as a dual American-Canadian citizen. Voting from Canada has not generally presented difficulties beyond timing, as the absentee ballot must arrive at the local election office by the close of polls on Election Day. Limits on when a voter could apply for an absentee ballot made for a potentially tight timeline for requesting, receiving, and submitting a ballot posted through international mail, even from Canada. The process for requesting a ballot in Kentucky has become more fully electronic in recent years, with the option to receive it by email and print it at home, easing the time pressure. Living in Windsor has often allowed me to cross the border to Detroit and mail my ballot using the U.S. Postal Service, avoiding international mail delivery costs and time lags. While this is an anomaly of living in a border city, crossing the U.S.–Canada border to mail my ballot was not possible in 2020 because it was closed to non-essential travel, and mail delivery between the two countries slowed considerably. Indeed, my primary election ballot in June 2020 made it to its destination in the county election office two weeks late and four weeks after I posted it from Canada, and it crossed the border three times before delivery. While it is difficult to know how many extraterritorial American voters faced the same issue in 2020, my general election ballot did arrive in Kentucky on time and with no difficulties. My vote for Joe Biden, however, counted for little in a state the incumbent president won by 26 percentage points.

This raises the question of why I would vote at all, pointing back to Kull's (2008) arguments about interests, community, and location in extraterritorial voting. I do not live in Kentucky, have no plans to return there, and knew in advance through polling that Donald Trump would almost certainly win the state's Electoral College votes. I voted for three reasons: a sense of civic obligation; a belief that as long as my U.S. tax filing obligations and access to absentee ballots and passport renewal remain in the hands of legislators over which I have some electoral power, I should exercise my right to vote; and finally, to be fully honest, a personal dislike for Donald Trump. I was not alone among American voters in Canada, as local media outlets in Windsor, Vancouver, and the Niagara region all reported in the weeks prior to the election that U.S. voters in Canada were requesting ballots in record numbers and, in a tight contest, could tip the election outcome (Barker 2020; Ryan 2020; Walter 2020). The focus on Windsor-based American voters was especially strong because of the dense networks of social, economic, and political connections across the Windsor-Detroit border region. Many of these voters cast their ballots in Michigan, which Donald Trump won in 2016 by a narrow margin, and where polling suggested a potentially tight 2020 contest (Barker 2020; Georgieva 2020). The border closure, discussed below, impacted the region strongly and became an election issue alongside national and local governments' pandemic response on both sides of the border. A late October editorial in the *Windsor Star* newspaper even exhorted Americans in Windsor and farther afield to vote against Trump "in great and defining numbers" (Jarvis 2020).

Voter mobilization organizations also targeted eligible Americans in the Niagara region, many of whom vote in neighboring New York, citing the border closure, poor U.S. pandemic response under Trump, and other American policies as directly impacting Canada and quality of life in U.S.–Canada border regions where many Americans live and work (Walter 2020). Barack Obama's ambassador to Canada, Bruce Heyman, told a Toronto audience that the 2020 election represented "an existential decision" for Americans, including those living abroad, while James Blanchard, U.S. ambassador to Canada under Bill Clinton, stated that the votes of Americans abroad were especially important and that the election would "determine who we are as a people and who we are in relation to the world, including our most trusted partner, Canada" (quoted in Keenan 2020). In the local Canadian contexts noted here, the common thread was strong mobilization among Democratic voters and by local chapters of the group Democrats Abroad (DA), which even ran ads on buses and local media in border cities in Ontario (Barker 2020). Non-partisan groups such as the Overseas Vote Foundation and Vote From Abroad also provided information on registration and balloting to American voters in Canada and other countries. These efforts opened and strengthened the transnational political space in which American voters in Canada define themselves as expatriates, emigrants, and dual citizens. Like the example of extraterritorial French voters and Macron's En Marche! party and movement, however, this space remains inchoate and its "normalization" incomplete because it is unclear if many of those participating in the 2020 election will continue to engage as part of a transnational political community of expatriate American voters beyond the threat of a second Trump administration.

The institutions that might give this transnational space and community form and durability are also a question, as the U.S. party system only weakly organizes and mobilizes extraterritorial American voters. With low turnout, averaging around 7 percent of eligible voters globally and often under 5 percent in Canada (Serhan 2020), and no specific representation for expatriate Americans, there is little added benefit to securing the votes of Americans abroad. How the Democrats and Republicans approach extraterritorial voters as a bloc also matters. While DA actively seeks to register and mobilize U.S. voters outside the country, even holding a "global primary" and sending delegates to the Democratic National Convention, Republicans Overseas (RO) is formally structured as a non-profit group and focuses mostly on fundraising and advocacy on issues relevant to Americans abroad, especially tax reform (Kalu and Scarrow 2020). The previous role of political parties in helping overseas voters obtain and return ballots in a timely fashion has been somewhat superseded by the Federal Voter Assistance Program (FVAP). Housed in the Department of Defense since 2009 and with the mandate to ensure that American military members, dependents, and civilians living abroad have reasonable access to and ability to vote in U.S. elections, FVAP is the primary federal government agency enacting the 1986 Uniformed and Overseas Citizens Absentee Voting Act (UOCAVA) and the 2009 Military and Overseas Voter Empowerment (MOVE) Act, which establish federal guidelines for American citizens' voting rights outside the U.S. (FVAP n.d.).

Kalu and Scarrow (2020, p. 891) suggest that U.S. "parties still have a potential role to play in reminding citizens about their voting rights, and in encouraging them to exercise them." In 2020, DA and RO chapters in Canada focused primarily on getting out the vote, as they already appeal to party-aligned voters and have limited resources to do traditional campaigning that also targets undecided voters. Kalu and Scarrow (2020) note that the number of country-level DA committees abroad has increased from 21 in 2000, to 54 in 2019. Steve Nardi, the chair of Democrats Abroad in Canada, stated that the group's Canadian membership had grown by 73 percent during Trump's presidency, with new local chapters opening in Windsor, the Niagara region, and the Atlantic provinces, and 400 volunteers working across Canada to mobilize expatriate Democratic voters to cast their ballots on time in 2020 (Purdon and Palleja 2020). Data that DA chapters collect about ballot requests suggests turnout among American voters in Canada was much higher than the historic norm of about 5 percent, but until FVAP provides more conclusive survey results of these voters' 2020 turnout, it is hard to know how many did indeed cast a ballot.

Despite media attention on American voters in Canada, their potential impact on the 2020 election, and efforts by groups like DA to mobilize them, neither presidential candidate had much to say about specific issues facing American voters in Canada or living anywhere else abroad. Democratic primary candidate Bernie Sanders came to Canada in July 2019, but to address healthcare issues and disparities as part of his campaign rather than to court American voters living there, highlighting the much lower cost of insulin in Canada as compared to the U.S. in front of a pharmacy in Windsor's Walkerville neighborhood (Charlton 2019). For both major political parties and presidential campaigns in the U.S., Canada remained a separate self-contained political entity and a mirror for U.S. policy discussions rather than the home to tens of thousands of potential voters who could sway the election. The border closure did charge this dynamic with some tension, however, not just for American voters in Canada but more broadly as part of the U.S.–Canada bilateral relationship, as discussed in the next section.

Border Politics and the Windsor-Detroit Region

Typically, the bilateral relationship with Canada does not register as a major issue in American elections. Trade disputes, such as those over softwood lumber or dairy, or the broader context of U.S.–Mexico–Canada economic integration and free trade agreements, are often the primary reason Canada has registered with American candidates and voters at all, and the two countries remain tightly bound by social and cultural affinities in addition to deep economic ties. As president, however, Donald Trump often made bombastic remarks about Canada and Liberal Prime Minister Justin Trudeau; imposed, rescinded, and then reimposed tariffs on Canadian aluminum imports; and threatened tariffs of up to 25 percent on auto imports, which likely would have decimated the auto sector in the Detroit-Windsor cross-border region (Automotive News Canada 2020; Gillies 2020). While Biden also posed challenges to the U.S.–Canada relationship because of his campaign promise to end the cross-border Keystone XL pipeline project, Trump's economic saber-rattling against the U.S.'s closest trading partner and divergence from Canada on several foreign policy fronts made him deeply unpopular among Canadians. Polls in summer and fall 2020 indicated that more than 80 percent of Canadians would vote for Biden given the chance, though approximately 40 percent of Canadians identifying as Conservative said they preferred Trump (de Clercy et al. 2020; Fournier 2020).

The pandemic heightened these tensions as the U.S. and Canada agreed to temporarily close the border to all non-essential traffic in March 2020 to control the spread of coronavirus. This mutual agreement was reviewed and extended monthly for more than a year and a half until some entry requirements began to be relaxed in late 2021. In summer and fall 2020, the border closure became an election issue in the U.S., wrapped up in the persistent trade battles Trump provoked with Canada and more fundamental concerns and rhetoric about the security of U.S. national borders. Because the border closure was mutual, enforced from the Canadian side as well, it also threatened the image of strength, negotiating prowess, and international weight that Trump attempted to project. By fall 2020, as a second and more deadly wave of Covid-19 began in the U.S. and Canada, Trump insisted on the campaign trail that the border would re-open "soon" and that Canadians and Americans both wanted the border open (Harris 2020; Zimonjic et al. 2020). Yet neither of these points was accurate. Polls conducted throughout 2020 showed Canadians overwhelmingly wanted the border to remain closed due to concerns about Covid-19 spreading from the U.S., and the U.S.–Canada land border remained closed to non-essential travel until Canada partially opened to vaccinated Americans in August 2021.

Nowhere was the border closure more impactful than the Detroit-Windsor region, where tensions immediately arose around truck traffic, healthcare workers, and broken family ties in a tightly-knit cross-border region with a deeply integrated economy (Darroch et al. 2020; Heuton 2020; Wilhelm and Reindl 2018; Workforce Windsor-Essex 2017). With U.S.–Canada trade tensions high throughout Trump's term in office, the border's Covid-related closure disrupted an already-unsettled situation for the thousands of trucks and truck drivers that cross the border daily. Not long after the border closure, the Canadian federal government quickly declared truck drivers essential workers and made them exempt from the 14-day quarantine rule and testing requirements that other travelers faced, and in the 11 months following the implementation of border restrictions, almost half of all entries into Canada were by truck drivers (Harris 2021). The first few months of restrictions still saw steep declines in trade volume across the border, especially in the Detroit-Windsor region, where cross-border truck traffic in April 2020 was down more than 50 percent from the normal daily count of 8–10,000 trucks and Cdn\$500 million of goods, largely due to widespread shutdowns in the auto industry (Waddell 2020). With non-essential travel almost completely halted, crossings by private car were down 70 percent at the Ambassador Bridge, the busiest border crossing in North America, and 85 percent at the Windsor-Detroit Tunnel in June 2020 (Darroch et al. 2020).

Most of those crossing by car at these two points of entry since the pandemic began are the approximately 1600 essential healthcare workers who live in Windsor but work in Detroit-area hospitals. As Covid rates surged across Michigan in spring 2020, federal and provincial authorities in Ontario instructed nurses and doctors who held healthcare jobs at institutions on both sides of the border that they had to pick one side to avoid being a vector for Covid to enter Canada from the U.S. The cross-border healthcare sector became a flashpoint in local and provincial debates over pandemic control measures in Windsor and other Ontario border cities such as Sault Ste. Marie and Sarnia, and a bargaining chip between Trudeau and Trump in discussions about trade and the border closure leading into the U.S. election cycle (Bouffard 2020; CBS News 2021; Weeks 2020). When Trump attempted in March and April 2020 to prevent 3M from exporting masks and respirators to foreign markets, including Canada, Trudeau threatened retaliation, including preventing healthcare workers from crossing into the U.S. (Austen 2020; Forrest 2020). While tensions eased somewhat after this, the heightened rhetoric around the border closure made the stakes very high for both countries, for local, state, and provincial officials along the border, for nurses and other essential workers navigating new border restrictions, and for Trump's re-election campaign.

The election made the usually technical aspects of U.S.–Canada border management more salient and immediate, while also demonstrating both federal governments' limited attention to region-specific border issues beyond their position in trade talks and political posturing relative to domestic audiences. All the while, Detroit-Windsor remained divided by the closure, with family members separated by a few miles unable to see each other for over a year. Meanwhile, the image of Canada as the stable "apartment above a meth lab" became fodder for election season jokes and memes just as tensions around the border closure grew and Trump and Trudeau traded economic threats. While Biden struck a different chord with his Canadian counterpart after taking office, talks between the U.S. and Canadian governments on how, when, and whether to reopen the border were practically non-existent well into 2021.

In this context, Canada again served as a backdrop and mirror for many American voters who articulate a Democratic or Republican partisan position by, respectively, threatening to move to Canada if Trump won, or casting Canada as a socialist society defined by healthcare rationing and high taxes. Neither of these positions were new in 2020, however, as Motyl (2014) examined similar rhetoric around the 2012 presidential election, arguing that threats of emigration are mediated by an eroding sense of political belonging following elections. The threat of "moving to Canada" following a preferred presidential candidate's loss is so widespread that *Forbes* even ran a column in late September 2020 by an American immigration lawyer providing tips on how to emigrate to Canada (Semotiuk 2020). Canada's conservative-leaning *National Post* newspaper, meanwhile, reported late in the election cycle that Canadians who had long lived in the U.S. were considering a return "home" if Trump won, with one even noting that some American friends had suggested marriage as an exit strategy to Canada (McCarten 2020). All of this ignores, however, a third and final trend related to the 2020 U.S. election in Canada, namely, how Trump-style populist and nationalist rhetoric has merged with existing forms of Canadian nationalism to help mainstream the far right's style and presence in Canadian politics.

"Take Back Canada" and Far-right Politics in Canada

The lessons of the bombastic Trump presidency and campaign have not been lost on Canadian politicians, activists, and observers. Erin O'Toole, member of Parliament for the suburban Toronto riding of Durham and a former Minister for Veterans Affairs, won leadership of Canada's Conservative Party in August 2020 with a campaign based on the slogan "Take Back Canada." Without ever directly stating from whom Canada should be taken back, O'Toole dabbled in Trump-style rhetoric throughout his leadership campaign, in his platform document, and in the role of official opposition leader in Parliament. In his platform, O'Toole argued for "Canada First" economic policies and more stringent immigration and asylum policies, targeted Justin Trudeau's "cowering approach to the Communist Party of China," and lamented that Liberal leadership had made Canada "complacent and naïve," while promising to bring "real world experience" to government to "take on the elites and rattle the system" (O'Toole 2020, pp. 6–7, 41). This echoed Trump's repeated claims about American failures and weakness, the role of unnamed "elites" in undermining American prosperity, and the threat of China, leftist ideology, and internal challenges to social order.

Commentators and researchers in Canada have been quick to point out how O'Toole and other federal Conservatives have drawn on Trump's belligerent rhetoric and political playbook, particularly in deploying economic nationalism to court White suburban and working-class voters, just as Trump made gains with similar voters across the U.S. Rustbelt in 2016 (Carlaw 2021; Savage and Black 2020). Former Conservative Party leadership hopeful Maxime Bernier, billing himself as "the Albertan from Quebec," broke from the party in 2018 to establish a new People's Party of Canada (PPC) on a platform of small government and strict immigration restrictions. Though the new party attracted many who espoused Trump-style populism and economic nationalism, Bernier himself did not fully leave his libertarian roots to embrace these political standpoints. The PPC thus had minimal impact on the 2019 Canadian federal election, winning no seats and less than 2 percent of the national vote (Gillies 2021). While the PPC increased this to almost 5 percent of the vote in the September 2021 federal election, the party still won no seats in Parliament. Others have pointed to similarities between Trump and provincial-level political trends and leaders, especially in Alberta, Ontario, and Quebec (for representative examples, see Benzie 2021; Harrison 2020; Authier 2016; Fournier 2021).

Yet current claims of "Trumpism" making inroads at the federal and provincial levels, especially but not only within the Conservative Party, must be taken with caveats. Many of the trends pursued by conservative political parties at both levels long predate the 2020 election cycle and Trump's own ascension in American politics. The populist rhetoric that conservative Canadian politicians and parties have adopted in the Trump era follows closely what Judis (2016, p. 15) identifies as "rightwing populism," which is "triadic," aiming upward at an "establishment" of cultural, economic, and political elites "but also down upon an out group" that is positioned as unfairly benefiting from elite favor. Judis (2016, p. 15) counterposes this against a "dyadic" leftwing populism which offers "a vertical politics of the bottom and middle arrayed against the top" and seeking various forms of solidarity or intersectional coalitions. But not all rightwing populisms are the same, and the Canadian context differs from the American in terms of immigration policy, policies and ideals of multiculturalism and national integration, and regional and provincial relations with the federal government. Examining these issues in detail is beyond the scope of this chapter but it suffices to say that if the 2016 election taught some Canadian conservatives how to hone their populist rhetoric, the 2020 election demonstrated Trumpism's limits in Canada, especially in relation to Trump's handling of Covid, widespread racial justice protests in the wake of George Floyd's murder, and his election loss.

Beyond the narrow confines of mainstream party politics and partisan leadership, however, the conditions that allowed Trump to win the presidency, kept him electorally competitive in 2020, and created a symbiotic relationship between Trump and fringe far-right and White nationalist movements also exist in Canada. Trump's November loss did little to undo these conditions, and the conspiracy-focused organizing that marked Trump's term, his 2020 campaign, and the larger reaction to Biden's election victory have found fertile ground in Canadian political soil. Canada's 2019 federal election even came under conspiratorial scrutiny in the weeks after Trump's own 2020 loss, based on false claims about the role of Dominion Voting Systems in counting votes that required Elections Canada to explain via social media that electronic vote tabulating machines are not used in Canadian federal elections (Press 2021).

Perry et al. (2019) also examine the surge in White nationalist organizing in Canada following Trump's 2016 election victory and an associated uptick in hate crimes against Black, Asian, and Muslim Canadians. They argue that Canadian manifestations of "the Trump Effect" are not surprising given "the economic, geopolitical and technological integration of Canada and the U.S." and existing domestic trends, including a long history of White nationalist and anti-immigrant organizing, at least two decades of conservative revanchism and neoliberal economic policies, and, most recently, Islamophobic rhetoric and policies through a decade of Conservative government led by Stephen Harper (Perry et al. 2019, p. 61). The rise of a right-wing media machine in Canada heavily favoring Trump and his political style, led by Rebel Media and far-right activist Ezra Levant, alongside forms of "digital vigilantism" on social media platforms, especially but not only in Quebec (Tanner and Campana 2020), have created a more permanent Canadian commentariat tied to Trump and the political far right, and integrated Canadian organizations and contexts into broader, more global networks of far-right figures and movements, including those beyond the U.S. and Trump himself.

Conclusion: The View from Canada Is also a View about Canada

In each of the three themes examined above – the American voter in Canada, the border as an election issue, and Trumpism in Canadian politics – the view of the U.S. election from Canada is thus also about Canada and Canadian society itself. In particular, the 2020 U.S. election presented a mirror for Canadians to address an uncertain future amid ongoing cross-border integration that suddenly seemed fragile and limited against the coronavirus pandemic, the border closure, and a surge of nationalist and populist rhetoric that heightened existing trends and contradictions in Canada. The durability and political value of Trump-style political rhetoric and policies in Canada following his November election defeat and his supporters' storming of the U.S. Capitol in January 2021 remains unclear, especially since, as noted, elements of an inchoate "Trumpism" were already present in Canada before Trump became politically relevant. Yet the emergence in Canada of political movements, parties, and figures echoing and referencing

Trump, his style, and his policies highlight how these can jump and shift between national contexts, punctuated but not solely defined by single electoral events.

The broad coalition of racial and social justice movements and organizations that emerged and often found common ground in the U.S. during the Trump presidency also found expression in Canada through the tumultuous summer of 2020 and into the fall U.S. election cycle. As with "Trumpism," however, these movements must adapt to the realities of Canadian society and politics or risk falling into political obscurity. The transnational political space in which American voters in Canada might create new political identities is also thus a space inhabited by diffuse networks of friends and family, conspiracy theorists, far-right groups, and populist and social movements of varying stripes, all adjusting to local and national contexts as they move back and forth between American and Canadian contexts. Geographers examining the 2020 election and its long tail will thus also have to contend with these forms of transnational political processes for a politically and geographically nuanced account of the election's impact and importance, and particularly the political fortunes of Trump and his form of rightwing populism in the U.S. and beyond.

Note

1 The original source for this comparison is the late actor Robin Williams, who made the joke during a 2013 "Ask Me Anything" event on reddit.com. Asked by a redditor if he would come to Canada soon, Williams replied that he loved Canada, that it was "the kindest country in the world," and that it was "like a really nice apartment over a meth lab." Emerging from the primordial digital soup on Reddit, the quip has since taken on a separate life as a sharable meme at convenient moments in the US–Canada relationship.

References

Austen, I. 2020. In Detroit she's a hero. In Canada she's seen as a potential risk. *New York Times* (April 10). https://www.nytimes.com/2020/04/10/world/canada/coronavirus-canada-detroit-nurses-hospital.html.

Authier, P. 2016. Analysis: François Legault zeros in on "elites" after Trump's upset in U.S. *Montreal Gazette* (Nov. 10). https://montrealgazette.com/news/quebec/analysis-francois-legault-zeros-in-on-elites-after-trumps-upset-in-u-s.

Automotive News Canada. 2020. Trump reimposes Canadian aluminum import tariff to stem "flood"; Canada retaliates. *Automotive News Canada* (Aug. 6). https://canada.autonews.com/trade-and-tariffs/trump-reimposes-canadian-aluminum-import-tariff-stem-flood-canada-retaliates.

Baringer, A., M. Herron, and D. Smith. 2020. Voting by mail and ballot rejection: Lessons from Florida for elections in the age of the coronavirus. *Election Law Journal* 19(3):289–320.

Barker, J. 2020. U.S. Democrats abroad keying in on Michiganders in southwestern Ontario as important vote. *CBC News* (Sept. 24). https://www.cbc.ca/news/canada/windsor/michigan-democrat-vote-2020-1.5736278.

Benzie, R. 2021. Doug Ford was Canada's most pro-Trump politician – until everything changed. *Toronto Star* (Jan. 20). https://www.thestar.com/politics/provincial/2021/01/20/doug-ford-was-canadas-most-pro-trump-politician-until-everything-changed.html.

Bergman, E. and P. Yates. 2011. Changing election methods: How does mandated vote-by-mail affect individual registrants? *Election Law Journal* 10(2):115–127.

Bouffard, K. 2020. COVID-19 tests Canadian relations as health staffers cross border to work in Detroit. *Detroit News* (April 19). https://www.detroitnews.com/story/news/local/michigan/2020/04/19/virus-tests-canadian-relations-health-workers-cross-border-detroit/5140321002/.

Burgess, K. and M. Tyburski. 2020. When parties go abroad: Explaining patterns of extraterritorial voting. *Electoral Studies* 66:102–169.

Carlaw, J. 2021. *Unity in Diversity? Neoconservative Multiculturalism and the Conservative Party of Canada* (Working Paper No. 2021/1). Toronto: Ryerson Centre for Immigration and Settlement and Canada Excellence Research Chair in Migration and Integration.

CBS News. 2021. Canadian nurses cross border to care for American patients: "It's not a border, just a line we cross." *CBS News* (June 5). https://www.cbsnews.com/news/nurses-cross-us-canada-border-care-for-american-patients/.

Charlton, L. 2019. U.S. presidential hopeful Bernie Sanders makes insulin bus trip to Windsor. *Windsor Star* (July 29). https://windsorstar.com/news/local-news/u-s-presidential-hopeful-bernie-sanders-makes-insulin-bus-trip-to-windsor.

Ciornei, I. and E. Østergard-Nielsen. 2020. Transnational turnout: Determinants of emigrant voting in home country elections. *Political Geography* 78:102–145.

Collyer, M. 2014. A geography of extra-territorial citizenship: Explanations of external voting. *Migration Studies* 2(1):55–72.

Darroch, M., R. Nelson, and L. Rodney. 2020. The Detroit-Windsor border and COVID-19. *Borders in Globalization Review* 2(1):42–45.

de Clercy, C., G. Seijts, and B. Nguyen. 2020. Do Canadian and American voters evaluate leader character similarly? Comparing voters' perceptions of Donald Trump, Hillary Clinton, and Justin Trudeau. *American Review of Canadian Studies* 50(4):498–521.

Dubin, J. and G. Kalsow. 1996. Comparing absentee and precinct voters: A view over time. *Political Behavior* 18(4):369–392.

Elul, G., S. Freeder, and J. Grumbach. 2017. The effect of mandatory mail ballot elections in California. *Election Law Journal* 16(3):397–415.

Federal Voter Assistance Program (FVAP). 2016. *Overseas Citizen Population Analysis Report, Volume 1: Participation and Voting Rates Estimation Prototype (February 2016)*. Arlington, VA: FVAP and Fors Marsh Group. https://www.fvap.gov/uploads/FVAP/Reports/FVAP-OCPA_201609_final.pdf.

Federal Voter Assistance Program (FVAP). 2018. *2016 Overseas Citizen Population Analysis Report (September 2018)*. Arlington, VA: FVAP and Fors Marsh Group. https://www.fvap.gov/uploads/FVAP/Reports/FVAP-2016-OCPA-FINAL-Report.pdf.

Federal Voter Assistance Program (FVAP). 2020. *2018 Overseas Citizen Population Analysis Report (July 2020)*. Arlington, VA: FVAP and Fors Marsh Group. https://www.fvap.gov/uploads/FVAP/Reports/2018-Overseas-Citizen-Population-Analysis-Report.pdf.

Federal Voter Assistance Program (FVAP). n.d.. About FVAP. https://www.fvap.gov/info/about.

Forrest, M. 2020. Trudeau warns U.S. against denying exports of medical supplies to Canada. *Politico* (April 3) https://www.politico.com/news/2020/04/03/3m-warns-of-white-house-order-to-stop-exporting-masks-to-canada-163060.

Fournier, P. 2020. How much do Canadians dislike Donald Trump? A lot. *Maclean's* (Oct. 1). https://www.macleans.ca/politics/how-much-do-canadians-dislike-donald-trump-a-lot/.

Fournier, P. 2021. Canada is not immune to Trumpism. *Maclean's* (Jan. 10). https://www.macleans.ca/politics/338canada-canada-is-not-immune-to-trumpism/.

Georgieva, K. 2020. Americans in Windsor area "on pins and needles" as election results trickle in. *CBC News Windsor* (Nov. 4). https://www.cbc.ca/news/canada/windsor/americans-in-windsor-area-pins-and-needles-election-results-1.5789002.

Gillies, J. 2021. The populist impact: The people's party and the Green Party. In J. Gillies, V. Raynauld, and A. Turcotte (eds.) *Political Marketing in the 2019 Canadian Federal Election*. pp. 75–90. Cham, Switzerland: Palgrave Pivot.

Gillies, R. 2020. Most Canadians hope for Trump defeat after insults, attacks. *AP News* (Oct. 30). https://apnews.com/article/donald-trump-virus-outbreak-toronto-global-trade-north-america-540a9b934c01b9571bf49b3c3513ce93.

Gimpel, J., J. Dyck, and D. Shaw. 2006. Location, knowledge and time pressures in the spatial structure of convenience voting. *Electoral Studies* 25(1):35–58.

Harris, S. 2020. Canada–U.S. border closure extended but Trump, Trudeau far apart on next steps. *CBC News* (Oct. 18). https://www.cbc.ca/news/business/trudeau-trump-canada-u-s-border-closure-1.5765323.

Harris, S. 2021. None of Ottawa's new travel rules apply to the largest group of people entering Canada – truckers. *CBC News* (Feb. 16). https://www.cbc.ca/news/business/non-essential-travel-truck-drivers-covid-19-test-border-1.5910888.

Harrison, T. 2020. Opinion: Trump-style politics the new normal in Alberta. *Edmonton Journal* (Jan. 25). https://edmontonjournal.com/opinion/columnists/opinion-trump-style-politics-becoming-new-normal-in-alberta.

Hartmann, C. 2015. Expatriates as voters? The new dynamics of external voting in Sub-Saharan Africa. *Democratization* 22(5):906–926.

Heuton, R. 2020. Windsor: An international border city in Detroit's shadow. In D. Cepiku, S. Jeon, and D. Jesuit (eds.) *Collaborative Governance for Local Economic Development: Lessons from Countries around the World*. pp. 30–48. Abingdon, UK: Routledge.

Jaca, G. and A. Torneo. 2020. Explaining (non) participation in overseas voting: The case of overseas Filipino voters in Japan in the 2016 elections. *Diaspora Studies* 14(1):45–74.

Jarvis, A. 2020. Jarvis: To our American friends, please vote, in great and defining numbers. *Windsor Star* (Oct. 21). https://windsorstar.com/news/local-news/jarvis-to-our-american-friends-please-vote-in-great-and-defining-numbers.

Judis, J. 2016. *The Populist Explosion: How the Great Recession Transformed American and European Politics*. New York: Columbia Global Reports.

Kalu, V. and S. Scarrow. 2020. U.S. parties abroad: Partisan mobilising in a federal context. *Parliamentary Affairs* 73(4):887–900.

Karp, J. and S. Banducci. 2000. Going postal: How all-mail elections influence turnout. *Political Behavior* 22(3):223–239.

Keenan, E. 2020. U.S. get-out-the-vote efforts reaches out to Canadian residents – and Donald Trump's fate might be at stake. *Toronto Star* (Sept. 5). https://www.thestar.com/news/world/2020/09/05/us-get-out-the-vote-efforts-reaches-out-to-canadian-residents-and-donald-trumps-fate-might-be-at-stake.html.

Kernalegenn, T. and C. Pellen. 2020. En Marche, French expatriates! The booming emergence of a new political actor among French residents overseas in the 2017 elections. In T. Kernalegenn and É. Van Haute (eds.) *Political Parties Abroad: A New Arena for Party Politics*. pp. 96–114. Abingdon, UK: Routledge.

Kull, C. 2008. Who should vote where? Geography and fairness in migrant voting rights. *Geographical Research* 46(4):459–465.

Leal, D., B.-J. Lee, and J. McCann. 2012. Transnational absentee voting in the 2006 Mexican presidential election: The roots of participation. *Electoral Studies* 31(3):540–549.

McCarten, J. 2020. "If Trump wins again, I'm moving": For some Canadian expats, option to return to Canada is a safety net. *National Post* (Oct. 23). https://nationalpost. com/news/world/if-trump-wins-again-im-moving-for-some-canadians-expats-option-to-return-to-canada-is-a-safety-net.

Minnite, L. 2011. *The Myth of Voter Fraud*. Ithaca and London: Cornell University Press.

Motyl, M. 2014. "If he wins, I'm moving to Canada": Ideological migration threats following the 2012 U.S. presidential election. *Analyses of Social Issues and Public Policy* 14(1):123–136.

O'Toole, E. 2020. Our country: A call to take back Canada. https://www.macleans.ca/ wp-content/uploads/2020/06/Erin-OToole-OurCountry-EN.pdf.

Perry, B., T. Mirrlees, and R. Scrivens. 2019. The dangers of porous borders: The "Trump effect" in Canada. *Journal of Hate Studies* 14(1):53–76.

Porter Robbins, C. 2020. https://www.theglobeandmail.com/world/article-these-americans-living-in-canada-agree-that-voting-in-the-us/.

Press, J. 2021. "This is unfortunate": Inside elections Canada after Trump's tweet on voting machines. *CBC News* (Jan. 3). https://www.cbc.ca/news/politics/elections-canada-trump-voting-machines-1.5860054.

Purdon, N. and L. Palleja. 2020. Why Republicans and Democrats are fighting a U.S. presidential election campaign battle in Canada. *CBC News* (Oct. 11). https:// www.cbc.ca/news/canada/national-republican-democrat-voters-in-canada-1. 5303171.

Rashkova, E. and S. Van Der Staak. 2020. Globalisation and the movement of people: What It means for party politics? An introduction. *Parliamentary Affairs* 73(4):831–838.

Ryan, D. 2020. "The interest is widespread on every side": Americans in Canada voting in unprecedented numbers. *Vancouver Sun* (Nov. 2). https://vancouversun.com/ news/local-news/americans-in-canada-voting-in-unprecedented-numbers.

Savage, L., and S. Black. 2020. How Erin O'Toole's strategy to win over union voters could work. *The Conversation* (Sept. 15). http://theconversation.com/how-erin-otoolesstrategy-to-win-over-union-voters-could-work-146259.

Semotiuk, A. 2020. How to move to Canada if Trump gets reelected. *Forbes* (Sept. 29). https://www.forbes.com/sites/andyjsemotiuk/2020/09/29/how-to-move-to-canada-if-trump-gets-reelected/?sh=1c6aab692ac0.

Serhan, Y. 2020. The absent voting bloc that could decide the U.S. election. *The Atlantic* (Oct. 30). https://www.theatlantic.com/international/archive/2020/10/ overseas-americans-could-help-decide-us-election/616888/.

Stewart, C. III. 2020. *How We Voted in 2020: A First Look at the Survey of the Performance of American Elections*. Cambridge, MA: MIT Election Data and Science Lab.

Tager, M. 2006. Expatriates and elections. *Diaspora* 15(1):35–60.

Tanner, S. and A. Campana. 2020. "Watchful citizens" and digital vigilantism: A case study of the far right in Quebec. *Global Crime* 21(3–4):262–282.

Waddell, D. 2020. Transportation sector and cross-border traffic slow by more than half. *The Windsor Star* (April 16). https://windsorstar.com/news/local-news/transportation-sector-and-cross-border-traffic-slow-by-more-than-half.

Walter, K. 2020. Niagara Democrats abroad urging Americans in Canada to "vote now". *St. Catharine's Standard* (Aug. 23). https://www.stcatharinesstandard.ca/news/niagara-region/2020/08/22/niagara-democrats-abroad-urging-americans-in-canada-to-vote-now.html.

Weeks, C. 2020. Ontario hospitals ban health-care workers from criss-crossing borders to work. *Globe and Mail* (April 6). https://www.theglobeandmail.com/canada/article-ontario-cities-ban-health-care-workers-from-crossing-borders-to-work/.

Wilhelm, T. and J. Reindl. 2018. Cross-border bond: Windsor-Detroit history of culture and commerce. *Windsor Star* (June 22). https://windsorstar.com/news/local-news/cross-border-bond-windsor-detroit-history-of-culture-and-commerce-could-be-tested-with-trade-war.

Workforce WindsorEssex. 2017. *Cross-Border Employment in the Windsor-Essex and Southeastern Michigan Corridor*. Windsor, ON: Workforce WindsorEssex.

Zimonjic, P., K. Simpson, and A. Panetta. 2020. Trump says Canada wants to reopen the border. But do we, really? *CBC News* (Sept. 18). https://www.cbc.ca/news/politics/trump-border-reopen-canada-1.5730806.

10 Fear, Joy, and Socialism in Cuban and Cuban-American Perspectives of the 2020 Presidential Election

John Paul Henry and Abraham Stephenson

In this chapter we analyze political rhetoric and affectual intensities concerning Cuba and Cuban-Americans during the 2020 U.S. presidential election. We argue that both political rhetoric and new media during this period follow the turn to post-truth, especially or the U.S. political right and Cubans opposed to ending the U.S. embargo. This political discursive divide had material consequences in Miami-Dade space-making performances, including voting. Our focus is on Cuban and Cuban-American narratives toward U.S. politics, specifically, as a full examination of geopolitical narratives concerning Cuban relationships with other countries is beyond the scope of this chapter.

This chapter is organized as follows. First, we elaborate general Cuban political positions toward U.S. politics, which are heavily influenced by embargo narratives. We then provide an analysis of political rhetoric concerning Cuba made by U.S. politicians during the 2020 campaign. We then analyze the rhetoric of prominent Cuban and Cuban-American activists. We follow this with a change of methods and analyze how affectual media reverberates through new media, relative to the Biden and Trump campaigns.

Cuban Positions on U.S. Politics

Cuban views of U.S. politics are deeply entangled with the U.S. embargo. This context is important as it frames many political narratives: by Cubans, by the Cuban state, and by Republicans. The embargo is seen as a tool to fight communism by the political right while it is often construed as a human rights abuse by the left. In this section, we provide an overview of the U.S. embargo against Cuba. We follow this overview with examples of how the embargo is discursively framed.

Following the success of the revolution led by Fidel Castro and the ousting of U.S.-backed Cuban president Fulgencio Batista on January 1, 1959, the revolutionary government began nationalizing farmland and private business. Economic tit-for-tat escalated in the following months, culminating in the failed U.S.-backed invasion of Cuba by Cuban exiles. On October 13, 1961, the United States instated an "economic embargo on Cuba, a ban on all US exports except medicines and some foodstuffs" (Pérez 2003, p. 243).

DOI: 10.4324/9781003260837-10

U.S. economic policies of the 1990s further attempted to strangle the Cuban regime following the fall of the Soviet Union. The Torricelli Bill in 1992

> prohibited subsidiaries of US corporation in Third World countries from trading with Cuba. Other features of the law authorized the president to withhold US foreign aid, debt relief, and free trade agreements with countries that provided assistance to Cuba. All ships trading with Cuba were denied access to US port facilities for a period of 180 days after having visited the island. Even before the Torricelli Bill passed, President George Bush issued an executive order banning all ships trading with Cuba from making port calls into the United States
>
> (Pérez 2003, p. 263)

Furthermore, the Helms–Burton Act of 1996 was intended to force other nation-states into stopping trade with Cuba. "The bill threatened with lawsuits any foreign companies that 'trafficked' in property previously owned by the United States. Foreign executives of those companies, and members of their immediate family, were also to be denied visas to enter or study in the United States" (Pérez 2003, p. 270). The extent to which U.S. foreign policy has been crafted to cause economic harm to the Cuban state has clearly become a focal point of Cuban national discourse. In the following section we show how the embargo has been framed in Cuban perspectives of U.S. elections.

Ending the Embargo

The most visible discourse of the Cuban state is the condemnation of the U.S. embargo. Through state-controlled media, the Cuban state conveys that a good U.S. president would be in favor of lifting the embargo. The state media discredited President Trump for activating Title III of the Helms–Burton Act (Fowler III et al. 2019; Isla 2021), reducing remittances, and restricting American travel to and from Cuba. The Cuban state understandably supports presidencies that would soften economic restrictions.

During the 2020 election, many Cubans showed support for Democratic candidate Joe Biden because they thought Biden would follow the same economic policies as President Barack Obama. Working-class Cubans, and especially *cubanos de a pie*, want change. *Cubanos de a pie* are people without transportation, who toil for minuscule state pay or are unemployed, and waste countless hours in ration lines. Most Cubans who live on the island want to see transformations so that there is food, good wages, clothing, footwear, and leisure. The July 11, 2021, protests (known as 11J) are an embodiment of this desire.

A second group of Cubans hold the position that the embargo should be lifted because it is an aggressive policy toward the country (Canal Caribe 2019). This includes many inside Cuba. In their analysis of Cuban political discourse on blogging platforms, Vicari (2014, p. 1011) found that state-aligned blogs such as *La Joven Cuba* framed political issues in terms of identity where political actors "remain behavioural icons and participate in the reiteration of the traditional revolutionary vocabular of action." In other words, many Cubans are reiterating state narratives concerning the embargo.

However, there is a part of the Cuban political opposition that advocates the lifting of the embargo. The opposition's argument is that the embargo is the excuse the Cuban government uses to justify its economic incompetence. Cuban activist and renowned journalist Yoani Sánchez takes this position. Sánchez has argued that the "'five-decade prolongation of the 'blockade' has allowed every setback we've suffered to be explained as stemming from it, justified by its effects'" (Betancourt 2014, p. 182). Sánchez is referring to how the Cuban state uses the embargo as a scapegoat for Cuba's economic malaise (BBC News Mundo 2013). In other words, by ending the embargo, the communist regime will no longer have an excuse for its economic failures.

Continuing the Embargo

Cuban activist Berta Soler takes the position that the embargo should be continued so as to apply pressure on the communist state. Soler "can be described as expressing strong support for the embargo as a signal of disapproval for the human rights abuses of the Cuban government" (Betancourt 2014, p. 183). Soler has likewise argued that the true embargo "está dentro de Cuba" or is located inside Cuba, a critique of Cuban state inefficiencies and corruption (Berta Soler 2013).

The pro-embargo political position is being silenced by two main practices. First, because this position is aligned with U.S. interests and against the Cuban state, it is simply not safe to repeat this political position. Consider the extent to which surveillance is conducted within Cuban society. Volunteering this political position would likely attract attention from the Committees for the Defense of the Revolution (Colomer 2000). Second, information about pro-embargo politics is being censored by the Cuban regime. For example, "Decreto Ley 370/2018, which went into effect on July 4, 2019, prohibits Cuban citizens and companies from using foreign servers to host websites or blogs. Chapter II of the decree law (articles 69 through 76) imposes fines for violations and allows the Ministry of Communications to confiscate and retain offending equipment" (Grant 2019, p. 4). This measure is intended to codify criminally practices used by independent journalists and activists for publishing information. However, *Decreto Ley 370* also regulates online speech and has been used to fine youth for "pre-delinquent" speech (*Multa de 2021*). *Decreto Ley 370* plays "a major role in the induction of self-censorship, by cultivating a sense of collective paranoia that causes citizens to limit the scope of their individual expression" (Herrera and Cañiva 2021, p. 103).

In other words, uneven access to information because of state censorship of the internet leads to disrupted political cognition. By political cognition we mean the "acquisition and elaboration of political knowledge" used for the "construction of interpretive schemata of events, actors and groups having to do with the political domain" (Vicari 2014, p. 1000). This uneven access to information is intensified racially. For example, a survey conducted by Hansing and Hoffmann (2020) of 1,049 Cubans found that 70% of Afro-Cubano respondents have no internet access whatsoever.

U.S. Political Rhetoric

In this section we discuss political rhetoric concerning the 2020 U.S. Presidential Election concerning Cuba and U.S.–Cuban relations. There is a clear difference between Democratic and Republican rhetoric. Democratic candidates tended to reference policies, while Republicans more readily turned to affective rhetoric, specifically referencing socialism. Our findings fit the "post-truth" pattern identified by Barfar (2019, p. 173) in which "objective facts are less influential in shaping public opinion than appeals to emotion and personal belief." In this section we analyze interview responses by both President Trump and President Biden to a popular independent Cuban news site. We then turn to politicians in Florida's 27th district, located within Miami-Dade County.

The Cuban independent news site *CiberCuba* published written interviews by both presidential candidates in the days immediately preceding the election. The candidates received similar questions concerning U.S.–Cuban policy, Cuban asylum and immigration, the sonic attacks on the American embassy, and remittances. These interviews, out of all the interviews conducted during the election year, are particularly important for its audience. *CiberCuba*, founded in Spain in 2014 by Cuban exiles, is an assemblage of editors, journalists, and collaborators located both within the island and the U.S. and Spain. According to the website traffic estimator *similarweb*, *CiberCuba* receives between two and four million visits each month, with almost 40% of traffic coming from the United States. In other words, an estimated 800,000 Spanish-speaking readers a month with concerns about Cuba turned to *CiberCuba* for information. While there are other Cuban independent news sites that garner more visits, like *PeriódicoCubano*, we focus on the interviews from *CiberCuba* for their direct engagement with the candidates.

Joe Biden

Published a day before the election in an interview with the Cuban independent news site *CiberCuba*, Biden said that his policy toward Cuba would be centered on two principles (*CiberCuba* 2020b). First, Biden said that Cuban-Americans are the best ambassadors for freedom in Cuba. Second, Biden said that empowering the Cuban people was central to the national security of the United States. To achieve these policy goals, Biden said he would eliminate restrictions on remittances and travel. Biden also said he would restore the Family Reunification Program (CFRP).

Biden attacked Trump's Migrant Protection Protocols, known as the "Remain in Mexico" policy. Under the "Remain in Mexico" policy, "non-Mexican asylum-seeks in the U.S. are sent to Mexican cities to await asylum hearings in U.S. immigration courts" (Chaparro 2020). Biden argued that under the Trump Administration there were nearly 10,000 Cubans "languishing in tent camps along the border with Mexico because of Trump's anti-immigrant agenda" (*Cubanos por El Mundo* 2020).

Donald Trump

President Trump responded to only five of the original twelve interview questions. Trump promoted his administration's sanctions against Cuba while attacking the Democratic Party by repeatedly associating Biden with socialism. For instance, Trump said that "As long as I am President, America will never be a socialist country! We need to defeat socialism and communism. I will not allow the United States to go the way of Cuba and Venezuela, which is where Joe Biden's radical agenda would take us" (*Entrevista* 2020). When asked why he has increasing support among the Cuban-American population, Trump said "Cubans have also experienced socialism and they don't like what they are seeing in the Democratic Party. What has happened to them? Before they were more normal. Now they are controlled by Fidel Castro-loving socialists like the madman Bernie Sanders, AOC [Alexandria Ocasio-Cortez] and Karen Bass, the communist" (*Entrevista* 2020).

Representative Shalala

Donna Shalala, a former U.S. Representative from Florida's 27th district, was born in Ohio, and has had a long public career. She worked in the Carter Administration, she served as U.S. Secretary of Health and Human Services during the Clinton Administration, and George W. Bush selected her to co-chair the Commission on Care for Returning Wounded Warriors. While not a Cuban-American, Shalala served as the president of the University of Miami from 2001 to 2015. In 2018 she was elected to the U.S. House of Representatives in Florida's 27th district, which is located entirely within Miami-Dade County, including Miami Beach and Little Havana.

In a post-election interview on November 6, 2020, Donna Shalala acknowledged the Democratic Party's failure to both message and act on topics important to Cubans and other Hispanics with first-hand experience of socialism. "We have to be very tough on Maduro, on what's going on in Cuba. Latin America is our best trading neighbor. We need to pay attention, much more attention than what democrats have been paying, to what's going on economically, socially, and politically," Shalala said (CBS Miami 2020). She went on to assert that "We have to demonstrate over and over again we are anti-communist. But you demonstrate that by really doing things and delivering, in addition to the messaging" (CBS Miami 2020).

Shalala said that the Trump Administration, while succeeding in messaging about socialism, has failed to act on Trump's promises. Shalala said "anti-communism, anti-socialism is deep within our community. But my argument is that Donald Trump hasn't delivered on any of that, he has just said it and scared people... Donald Trump accused us of being socialists. He didn't do anything about socialism and communism in Latin America. Latin America is not better off after four years of Donald Trump" Shalala said (CBS Miami 2020). This statement could easily be argued as the Trump Administration enacted a list of sanctions against the Cuban Communist Party.

Take, for instance, the Trump Administration's ending the suspension of the Helms–Burton Act on May 2, 2019. Title III of the Act "provides a private cause of action for U.S. nationals against any person or entity that knowingly 'traffics' in property expropriated by the Cuban government. The Act also provides exclusive jurisdiction for these claims in U.S. federal courts" (Fowler III et al. 2019). As of June 2021, Swiss multinational cement company LafargeHolcim agreed to settle a lawsuit filed in September of 2020 by William Claflin and 24 others. LafargeHolcim invested in the Cuban "Carlos Marx cement plant located on the confiscated property of Compañía Azucarera Soledad, SA, according to court documents" (Brasileiro 2021). The plaintiffs argue Compañía Azucarera Soledad was nationalized by the Cuban government on August 6, 1960.

Representative Salazar

Maria Elvira Salazar is a Cuban-American journalist turned politician. Salazar had a storied and successful journalism career before becoming a U.S. Representative. Salazar worked for CNN Español, covered the Salvadorian civil war, and interviewed Fidel Castro, according to her congressional website. She defeated incumbent Shalala in November 2020 to represent Florida's 27th Congressional District, which is located entirely within Miami-Dade County.

Salazar won the election in an upset against Shalala, *The Miami Herald* writes "by successfully tying Shalala to left-leaning Democrats like Vermont Sen. Bernie Sanders, a self-described democratic socialist" (Daugherty 2020). Indeed, her messaging about socialism is clear. For instance, on the "About" section Salazar's political website it reads "Maria Elvira Salazar is in Congress fighting to stop socialism from ever coming and ruining America" (*About* 2021).

Salazar is critical of the Biden Administration's Cuban policy, especially its lack of action after the 11J protests. Salazar introduced Operation Starfall in August after the Cuban government cut internet access to the entire population in an attempt to block the movement of sousveillant videos and other citizen journalism from leaving the island. "In September, the congresswoman requested that the House add Operation Starfall to the National Defense Authorization Act of 2022, but House Democrats blocked consideration of the plan" (*Internet* 2021). On November 3, 2021, the U.S. House of Representatives passed House Resolution 760 which urges the president to "expand internet access for the Cuban people" (*Internet* 2021).

Affective Rhetoric of Cuban and Cuban-American Activists

In this section we center the voices of prominent Cuban and Cuban-American activists. We start with members of the Group of 75, which consists of 75 Cuban activists and journalists arrested in the Black Spring of 2003. Narratives from selected activists highlight the struggle against communism and applaud efforts by U.S. politicians to diminish the Cuban Communist Party. We then turn to Cuban-American influencer and activist Alex Otaola, based in Miami, and his influence on the 2020 elections.

José Daniel Ferrer García

José Daniel Ferrer García, who lives in Santiago de Cuba, is one of the more influential and visible Cuban human rights activists. Ferrer was one of the 75 grassroots activists imprisoned by the state during the Black Spring of March and April 2003. Ferrer is the Coordinator General of the Patriotic Union of Cuba (UNPACU). Created by Ferrer in 2011, UNPACU has more than 3,000 activist members who, according to the organization website, practice activism based on resistance and non-violent disobedience. UNPACU has official representation in the U.S. and the European Union (*UNPACU* 2018).

Ferrer made abundantly clear his position on the Biden Administration. In the days leading up to the election, Ferrer wrote on Twitter that the administration would give the Cuban Communist Party money and "the ability to continue filling the US with communist agents while the regime destroys internal opposition" (Fernández 2020). Ferrer blamed the economic thaw with the Obama Administration for an increase in repression. On November 6, Ferrer published on Twitter the need for the United States to increase sanctions against a regime which enriches itself and represses citizens (Opositor 2020a). Ferrer was detained and imprisoned on July 11, 2021, for participating in the 11J protests (CiberCuba 2021a).

Berta Soler and Ángel Moya

Berta Soler is the leader of Las Damas de Blanco, or Ladies in White, which protests the political imprisonment of Cubans. Las Damas de Blanco formed in 2003 in response to the 75 journalists and activists arrested during the Black Spring. Every Sunday Las Damas de Blanco dress in white to "gather and attend mass wearing white, and then, march silently through the streets" (Borgen 2021). Soler lives in Lawton, a district of Havana.

Soler said that on February 2017, 2021, she delivered a letter, drafted by a group of Cuban activists, to a diplomat from the U.S. Embassy recommending policy to the new Biden Administration. The letter requests that U.S. concessions be made on the condition of:

> the freedom of all political prisoners; cessation of political repression, legalization of independent organizations; business freedom; free hiring of Cuban workers and direct payment of wages to workers; respect for the conventions of the International Labor Organization; Respect for human rights; and free, pluralistic, and uncoerced elections, with a transitional government that guarantees respect for the physical and moral integrity of political opponents and the participation of credible, reliable, and impartial observers from the UN, OAS, and the European Union.
>
> (*Carta* 2021)

Others associated with Las Damas de Blanco have taken to social media to support President Trump's stance against Cuba. Two members of Las Damas de Blanco, Maria Cristina Labrada Varona and Lourdes Esquivel Vieyto, support Trump on Facebook, saying "'Proud to be Cubans. If we could vote, we could vote for Donald Trump'" (Pentón 2020). Soler's husband, Ángel Moya, who was sentenced to 20 years imprisonment during the Black Spring but has since been released, also voiced his support for Trump's tough Cuban policy. "Whatever happens, Donald Trump leaves us a legacy of struggle and conduct against communism. We will defend his legacy!" (Authors' translation) (Resultados 2020).

Alex Otaola

Alex Otaola is a Cuban-American actor-turned-influencer and activist through his popular gossip and news program *Hola Ota-Ola!*, which streams live on YouTube. Otaola's influence on Miami politics in recent years should not be understated. Previously a Democrat, Otaola voted for both Obama and Clinton before renouncing the Democratic Party on air after the 2018 election of New York Congresswoman Alexandria Ocasio-Cortez, a Democratic Socialist (Mills 2020). Both Republicans and Democrats credit Otaola for a swell of Cuban-American support in south Florida (Mills 2020). Otaola's show airs on the online news site, *Cubanos por el Mundo*, which averages more than 250,000 monthly visits with more than 40 percent of visits coming from the U.S., according to the website traffic estimator *similarweb*.

Otaola leverages affect and disinformation to engage viewers. His show, *Hola Ota-Ola!*, should be understood as part of the post-truth turn in which "objective facts are less influential in shaping opinion than appeals to emotion and personal belief" (Barfar 2019). *Hola Ota-Ola!* is a commentary-style show which purposefully arouses anger in viewers (Hasell and Weeks 2016). The effects of Otaola's messaging are felt both in terms of Miami spaces and in voter turnout. Otaola met with and interviewed President Trump in October of 2020. However, during this meeting Otaola offered to give Trump a list of Cubans he said are tied to the Cuban Communist Party (Cubanos por el Mundo 2020). This list, however, has been critiqued as a form of neo-McCarthyism and an effort to intimidate Trump opponents and critics of Otaola (Taladrid 2020).

Otaola's activism affected Miami spaces. One of the clearest was the anti-communist caravan protesting relations with the Cuban dictatorship. The demonstration was held on Saturday, February 6, 2020. "The Cuban dictatorship and its acolytes intend to provoke us with pro-Castro caravans in Miami. Next week we will show who we are the majority. Anti-communist caravan and anti relations with the dictatorship, on Saturday, February 6. We are the red wall," Otaola wrote on his Facebook profile.

The threat of communism and socialism manifested in physical demonstrations of resistance in Miami. On October 10, 2020, thousands of people participated in a caravan billed as an Anti-Socialist and Anti-Communist Caravan with the goal of warning "the United States about the dangers of socialism" (Pentón 2020). Organizers estimate 20,000 people participated in the car rally, with a majority of participants carrying Trump signs. Alex Otaola helped organize the event as well. "We've had a massive response to this caravan. Many Hispanics, Cuban, Venezuelans, Nicaraguans are all here to say we reject communism and socialism and that we want the liberation of Cuba, Nicaragua, and Venezuela," said Orlando Gutiérrez-Boronat, Cuban Democratic Directorate in a video uploaded to the Miami Herald (Pentón 2020).

In the previous sections we demonstrated divergent political practices concerning Cuba coming from the U.S. political left and right. Specifically, we highlighted the affectual appeals in Cuban and Cuban-American political rhetoric. In the following section, we turn to new media representations and affectual amplifications of the 2020 election in Miami-Dade County.

New Media and Affect in Miami-Dade County

Finally, we turn toward the Cuban-American and Cuban diaspora in Miami-Dade County, Florida as an influential and motivated voting bloc with deeply personal and passionate geopolitical interests. This group is passionate because they have seen first-hand the realities of socialism and communism. This deep experiential knowledge spills over into political rhetoric, as seen previously in this chapter. While we identified the turn to post-truth rhetoric, there is a need to understand how affectual intensities are amplified in representations in new media spaces. This need is especially salient in key voting districts as these districts are inundated with political messaging. Therefore, we center the role of new media in amplifying affectual rhetoric and representations to understand political intensities communicated through social media during the final weeks of the 2020 U.S. election. We contextualize this need with the following research question: In what ways has a politics of affect been used relative to the Cuban-American population during the 2020 U.S. presidential election?

To answer this question, we first start by locating Miami-Dade County demographically. We then provide a brief history of Cuban waves of emigration to the United States. We follow this with an analysis of recent voting patterns in Miami-Dade County. Next, we go on to define affect and describe our methodology used to analyze political affectual intensities on Twitter as they relate to both Trump and Biden in the final weeks of the 2020 U.S. presidential election. We finish by providing a discussion on our findings of Cuban-American affectual intensities in Miami-Dade County.

Florida is a swing state with 29 electoral votes. Many of Florida's statewide elections are decided by only one or two percent margins (Gaffney 2020). The 2000 election perhaps encompasses this tendency toward narrow margins most clearly. President George W. Bush and Democratic candidate Vice President Al Gore both earned 48.8% of the popular vote (*2000 Results* 2021a). A 537-vote gap remained between the candidates. This difference swung the election to Bush, which had very real global consequences.

Narrow margins are seen in more recent elections as well, when voters in Florida twice supported President Obama before twice supporting President Trump. Floridians voted to elect Obama with 51% of the vote in 2008 compared with 48.2% for Senator John McCain (*2008 Results* 2021b). In 2012 Obama won Florida with a 0.9% margin against Mitt Romney (*2012 Results* 2012b). Then in 2016 Floridians voted to elect President Trump with a 1.2% margin over Secretary Hillary Rodham Clinton and again in 2020 with a 3.3% margin against President Biden (Gaffney 2020). No place in Florida is more important in influencing these elections than Miami-Dade County.

Located at the southern tip of the state, Miami-Dade County is the most populous county in Florida. In 2020 Miami-Dade County had a population of 2,701,767 (U.S. Census Bureau 2021). There were 1,563,572 registered voters in Miami-Dade County as of the 2020 election (*2020 General Election* 2020). There was a 74.59% turnout of 1,166,203 voters (*2020 General Election* 2020).

The racial composition of Miami-Dade County has changed dramatically in the last decade. In 2010 there were 1,856,938 people who identify as Hispanic or Latino, or 68.7% of the county population. According to the U.S. Census Bureau (2021), in 2020 there were 796,893 Miami-Dade County residents who identified as White, a 56.7% decrease from 2010.

Importantly, there are about 980,000 Cuban-Americans living in Miami-Dade County, or 36.3% of the population (Grenier and Lai 2020). Of this population about "70% of Cubans living in South Florida have relatives living in Cuba" (Grenier and Lai 2020, p. 24). This means that most Cuban-Americans in south Florida have a familial interest at stake when it comes to U.S.–Cuba geopolitics. Given Florida's status as a swing state, its significant number of 27 electoral votes, and the influence of Cuban-Americans living in Miami-Dade County, it is important to understand how political campaigns are leveraging affect in south Florida.

Cuban Emigration

It is first useful to understand the waves of Cuban emigration constituting much of the Cuban-American population before discussing the political views of Cubans in south Florida. We begin by discussing 19th-century immigration influenced by U.S. economic policies. We then follow this by discussing Cubans fleeing the island following the socialist revolution, followed by the period after the collapse of Soviet aid.

U.S. tariffs on the Cuban cigar industry in the 19th century precipitated first a mass relocation of cigar manufacturers and then immigration of the unemployed from Cuba to Florida (Pérez 2003). "Over the latter half of the nineteenth century, more than one hundred thousand men, women, and children, almost 10 percent of the population, took up residence abroad – in Europe, in Latin America, but especially and mostly in the United States" (Pérez 2003, p. 65). Waves of capital transfer from Cuba to Tampa were followed by Cubans seeking employment. Key West grew from twenty-eight hundred to eighteen thousand while "Tampa grew from two thousand to twenty-three thousand" (Pérez 2003, p. 66).

However, it was the fall of President Batista and the January 1, 1959, victory of the socialist revolution that would spur the most notable movement of Cubans from the island to Florida. Unable to successfully resist the revolution and dramatic economic changes from within, mostly middle-class, white, and educated Cubans fled the island (Pérez 2003). "The loss of population in the early years was stunning: sixty-two thousand in 1960, sixty-seven thousand in 1961, sixty-six thousand in 1962" (Pérez 2003, p. 245). Cuban emigration neared 1 million by the end of the 1980s while being supported by nearly $730 million in direct U.S. subsidies (Pérez 2003).

In the early to mid-1990s, Cubans, desperate to escape an array of dismal living conditions compounded by the collapse of Soviet aid, boarded makeshift rafts by the thousands to cross the Florida Straits. In response to the increasing *balsero* crisis, the Clinton Administration ended automatic asylum for Cubans intercepted in the Florida Straits, beginning the "wet foot, dry foot" policy. Under this policy, only Cubans who landed on U.S. territory would be granted asylum. The Obama Administration ended the "wet foot, dry foot" policy in January 2017, just before leaving office.

What Are Cuban-American Political Views in South Florida?

To describe the political identities of Cuban-Americans in Miami-Dade County, we rely on Grenier and Lai's (2020) Florida International University Cuba Poll. Conducted annually, this poll follows the pulse of Cuban-American politics. Following Grenier and Lai's (2020) estimate, there were about 482,306 Cuban-American voters in Miami-Dade County in 2020. In Miami-Dade County in 2020 an estimated 52.6% of Cubans are registered Republican while 21.5% are registered independent and 25.8 are registered Democrats (Grenier and Lai 2020).

A clear majority of Cuban-Americans in south Florida supported President Trump's policies in 2020. According to Grenier and Lai (2020), 64% of Cuban-Americans in south Florida supported President Trump's approach to immigration, while 66% supported his Cuba policy, 65% supported his handling of the Covid-19 pandemic, and 80% supported his approach to the economy. "Seventy-six percent of registered voters who arrived between 1010 and 2016 report registering as Republicans" (Grenier and Lai 2020, p. 4). In the same FIU Cuba Poll, a clear majority of Cubans support "policies designed to put maximum pressure on the Cuban government to promote regime change" (Grenier and Lai 2020, p. 17). Grenier and Lai (2020, p. 4) expected President Trump to receive 59% of the Cuban-American vote in 2020.

While Biden carried Miami-Dade County, the Democrats' lead shrunk significantly from that of Hillary Clinton. Biden received 617,864 votes or 53.31% while Trump received 532,833 or 45.98% (*2020 General Election* 2020). This is a dramatic shift from 2016 when Clinton won 63.22% compared to Trump's 33.83% (*2016 General Election* 2021). This voting swing is due in large part to consistent messaging, and disinformation, on the part of the Trump campaign to play to the fears of not only Cubans, but also the Hispanic population who immigrated from Venezuela, Colombia, and Nicaragua (Caputo 2020). As this chapter previously demonstrated, disinformation and affective rhetoric were used to attack Biden and Harris. Disinformation plays into fears and anxieties which were very personal to this population. In the following section, we discuss the capacity of new media to amplify these fears to targeted audiences in Miami-Dade County through new media.

Affect

Geographers have grown increasingly concerned for the implications of politics infused with affect (Sharp 2009), or affect program theory. Thrift (2008) cites Griffiths (1997, p. 79): "'In its modern form, the affect program theory deals with a range of emotions corresponding very roughly to the occurrent instances of the English terms surprise, fear, anger, disgust, contempt, sadness, and joy. The affect programs are short-term stereotypical responses involving facial expression, autonomic nervous system arousal, and other elements.'" "In other words, the affect programme approach makes a case for the view that certain so-called lower order affects at least have some degree of cultural generality but are not therefore necessarily innate. It makes no such claim for so-called higher cognitive affects such as love, jealousy, guilt and envy" (Thrift 2008, p. 224). Shaw and Warf (2009) demonstrated how affect is wielded through videogames, giving rise to these low-level responses. Attention to affect is crucial as it may show us the evolving mechanisms of political manipulation. An attention to affect cues us into how human geographies are stimulated with a press of a button. Politics of affect have spatial consequences, like the massive anti-communist caravans carried out in Miami in 2020.

Multimedia and new media, including Twitter and other social media sites, have been shown to illicit affectual responses. One component of multimedia is the visual. Pink (2013) cites Rose's (2003) argument that visual methodologies can show us "the ways in which particular visualities structure certain kind of geographic knowledges," especially in terms of power relationships. For instance, in his study of the Ohio River valley chemical corridor, Henry (2021) used visual research methods to critically assess how visuality affects perceptions of toxicity and place. Geboers and Van De Wiele (2020, p. 746) centered the affective capacity of images on Twitter concerning Syrian refugees "as key to understanding the intentional strategies for mobilization operating within an attention economy."

However, video also includes movement and aural components which offer more modalities to trigger affectual responses. Likewise, Mueller (2015, p. 417) notes that video offers "an immediate window on embodied and non-textual performances and addressing different sensory registers, including the visual and the aural." Research methods attuned to not only the visual but also aural and movement bring us closer to understanding place-based affectual intensities, as in the following case of Miami-Dade County.

Methods

In this section we will discuss how we analyzed affect relative to the politics of Cuban-Americans in south Florida viewed through Twitter. We begin with an overview of our content analysis method. We follow this with a discussion on our data source. We then follow this with a discussion of how we filtered tweets to obtain a spatially and thematically focused dataset.

We used content analysis (Rose 2016) to categorize base affects through visual messages related to the election. My framework also incorporates Rose's (2016) call for a critical method in which the site of production and the site of circulation are considered relative to other variables.

We used Hui's (2020) public domain dataset of 1.7 million tweets published between October 15, 2020 and November 8, 2020. The tweet data was scraped using the Python script snsscrape and the Twitter API using the hashtags #DonaldTrump, #Trump, #JoeBiden, and #Biden.

First, we selected only tweets originating from Florida. We then further filtered Florida tweets containing the terms "Cuba" and "http". Importantly, this filtering method includes both Spanish and English use of the word Cuban and Cubano. The "http" is relevant for identifying tweets containing hyperlinks to either photographs or website URLs. The resulting tweet selection overwhelmingly shows tweets originating from Miami. This filtering method returned 143 tweets. We then followed this same workflow for tweets concerning President Biden, returning 101 tweets originating in Florida. We then removed spam tweets and duplicate tweets which appeared in both the #Trump dataset and the #Biden dataset. Finally, we removed Biden victory tweets as they do not similarly represent the affective atmosphere during the campaign. The final dataset included 95 separate images and videos for individual analysis.

The variables we analyzed include the following: media type: video, landing page, image subject, image location, references to socialism (also communism and dictator), references to violence, the presence of Trump, the presence of Biden, references to sexuality, and representations of gender. Lastly, we noted whether the tweeted image elicited any of the following affects: surprise, fear, anger, disgust, contempt, sadness, and joy.

Results

Socialism

Pro-Trump tweets were twice as likely to reference socialism than were pro-Biden tweets. In the instances in which pro-Trump tweets referenced socialism the affective capacity was related to fear, anger, and contempt. Pro-Trump tweets used socialism as an emotional tool to attack Biden. This finding is expected. Consistent messaging by the Trump campaign appealed to fears of socialism (Caputo 2020). However, clearly pro-Trump media amplified the socialist content directly to Cuban-American viewers. For instance, YouTube influencer Alex Otaola, who has an estimated 100,000 viewers for each show, met with Trump on October 15, 2020 and repeatedly engages in critiques of socialism and Democrats (Mazzei 2020).

Traditional mass media sources, which tend toward the politically neutral, used socialism in relation to news stories. However, most often these news organizations were engaging in repeating campaign narratives. A recurring theme for both neutral sources and pro-Biden sources was the *rejection* of the socialist label.

Gender

In this section we will discuss the representation of gender in tweets originating from Miami. Women were represented in 40% of images containing people in the pro-Biden camp. Whereas women were represented in 26% of images containing people in the pro-Trump camp. However, *how* women were represented is important. In pro-Biden tweets, women were often engaged in political outreach or campaigning, compared to singing and dancing in pro-Trump tweets.

Affect

In this section we will discuss the affectual quality of images (e.g., giving rise to fear, anger, joy) and their relation to political allegiance. Pro-Trump tweets were decisively affectual when compared to the pro-Biden tweets, almost two to one. Of the pro-Biden tweets, 19% contained affective qualities. Meanwhile of the pro-Trump tweets about 42% had clear affective states. These data show a clear tendency for pro-Trump messaging to engage with a range of affectual energies. Curiously, the only affectual emotion *not* represented in the pro-Trump tweets is sadness.

Joy was the most frequently relied-upon emotion. Pro-Trump tweets account for 72.7% of the joyful tweets, while only 18.2% of pro-Biden tweets are joyful. The frequency was in part due to the amount of Trump rallies and caravans in Miami during the late campaign season. This reflects the actual presence the Trump campaign had in south Florida. A *Chicago Tribune* analysis reported that between September 1, 2020, and November 3, 2020, either Trump or Pence visited Florida a total of 17 times compared to the 13 visits by Biden or Harris (Ruthhart and Berlin 2020). Of these visits, however, the Trump campaign tended to hold rallies, while the Biden campaign tended toward drive-in rallies, roundtables, or voter-turnout events. In other words, the Trump campaign leveraged affectual-based events replete with dancing and live music. Joy for the Biden campaign, however, was lacking. Most Biden-related tweets tended toward the factual, rather than the affective. The most notable difference, however, is when Biden was called the winner. At this point joy was turned on and amplified in videos.

The Trump campaign effectively leveraged the affective power of music to amplify joy and literally move Cuban-Americans in Miami. The Miami-based musical group Los Tres de La Habana produced a voting anthem for the Trump campaign (*Tres* 2020). Released in October 2020, the catchy song refrains: "Oh my God! I will vote, I will vote for Donald Trump." Many tweets were published during the study period containing this tune in some way, amplifying a joyous atmosphere in the weeks preceding the election. After all, "affective attunement to music sustains moods and modes of engagement for individuals of varying ideological orientations, leading to both unifying effect and an impression of engagement" (Papacharissi 2015, p. 22). These findings show how the Trump campaign leveraged a politics of affect of both fear and joy in Florida's Miami-Dade County.

References

2012 Results. 2012a. 2012 Florida Presidential Results. *Politico*. https://www.politico.com/2012-election/results/president/florida/

2012 Results 2012b. Live Florida Election Results and Maps, includes 2012 races for President, Senate, House, Governor and Florida Ballot Measures Results. *Politico*. https://www.politico.com/2012-election/results/president/florida/

2020 General Election. 2020. MIAMI-DADE COUNTY Christina White, *Supervisor of Elections*. https://enr.electionsfl.org/DAD/2779/Summary/

About. 2021. Maria Elvira Salazar bio. *Maria Elvira Salazar for Congress*. https://mariaelvirasalazar.com/bio/

American Presidency Project. 2021a. 2000 | The American Presidency Project. https://www.presidency.ucsb.edu/statistics/elections/2000

American Presidency Project. 2021b. 2008 | The American Presidency Project. https://www.presidency.ucsb.edu/statistics/elections/2008

American Presidency Project. 2021c. *2000 Election Statistics*. https://www.presidency.ucsb.edu/statistics/elections/2000

American Presidency Project. 2021d. *2008 Election Statistics*. https://www.presidency.ucsb.edu/statistics/elections/2008

Barfar, A. 2019. Cognitive and affective responses to political disinformation in Facebook. *Computers in Human Behavior* 101:173–179.

BBC News Mundo. 2013. Yoani Sánchez recibe críticas y apoyos en Brasil. *BBC News Mundo*. (Feb. 19). https://www.youtube.com/watch?v=c1KBydf6l3M

Berta Soler. 2013. Berta Soler: "que no se levante el embargo ... el bloqueo a la Isla está dentro". (April 4). https://www.cubanet.org/actualidad-destacados/berta-soler-que-no-se-levante-el-embargo-el-bloqueo-a-la-isla-esta-dentro/

Betancourt, R. 2014. Should the US lift the Cuban embargo? Yes; it already has; and it depends! *Cuba in Transition 23*:175–185.

Brasileiro, A. 2021. In historic first, Swiss firm settles suit by U.S. family over seized property in Cuba. *Miami Herald* (June 1). https://www.miamiherald.com/news/nation-world/world/americas/cuba/article251823003.html

Canal Caribe. 2019. ¿Qué opinan los cubanos del Bloqueo de EEUU. contra Cuba? *Revista Buenos Días*. *Canal Caribe*. (Nov. 7). https://www.youtube.com/watch?v=qs--4xQj0Ho

Caputo, M. 2020. 'What do we do?': Trump gains rattle Miami Dems. *Politico*. (Sept. 8). https://www.politico.com/news/2020/09/08/trump-miami-florida-support-410362

Carta. 2021. *Berta Soler envía mensaje a Biden con un diplomático de EEUU (AUDIO)*. 2021 (Feb. 18). Radio y Televisión Martí | RadioTelevisionMarti.com. https://www.radiotelevisionmarti.com/a/berta-soler-envía-mensaje-a-biden-con-un-diplomático-de-eeuu/286096.html

CBS Miami. 2020. Facing South Florida: 1-On-1 with Donna Shalala. (Nov. 8). https://www.youtube.com/watch?v=aITBUe99SXA

Chaparro, L. 2020. While Cuban-Americans in Florida came out big for Trump, Cubans stuck in Mexico pinned their hopes on Biden. *Business Insider* (Dec. 1). https://www.businessinsider.com/cubans-stuck-in-mexico-because-of-trump-welcome-biden-election-2020-12

CiberCuba. 2020a (Nov. 7). *Opositor cubano José Daniel Ferrer publica mensaje al pueblo de EE.UU. y a sus representantes*. https://www.cibercuba.com/noticias/2020-11-07-u1-e129488-s27061-opositor-cubano-jose-daniel-ferrer-publica-mensaje-al-pueblo

CiberCuba. 2020b (Nov. 2) *Entrevista exclusiva a Joe Biden en CiberCuba*. https://www.cibercuba.com/noticias/2020-11-02-u1-e42839-s27061-joe-biden-limitaciones-remesas-solo-perjudican-familias-cubanas

CiberCuba. 2021a (Aug. 7). *Hija de Ferrer pide libertad para detenidos el 11J: No sabemos qué torturas les están haciendo*. https://www.cibercuba.com/noticias/2021-08-07-u1-e196568-s27061-hija-ferrer-pide-libertad-detenidos-11j-sabemos-torturas-les

CiberCuba. 2021b (Nov. 11). *María Elvira Salazar pide a Biden una estrategia para restaurar internet en Cuba si el régimen corta el servicio el 15N*. https://www.cibercuba.com/noticias/2021-11-11-u1-e129488-s27061-maria-elvira-salazar-pide-carta-joe-biden-este-listo-apoyar-al

Colomer, J. 2000. Watching neighbors: The Cuban model of social control. *Cuban Studies* 118–138.

Cubanos por el Mundo. 2020 (Oct. 17). *Entrevista de Alex Otaola a Trump: "Vas a tener libertad donde viniste, vas a poder ir y regresar."* https://www.youtube.com/watch?v=7rdRyLCxMxM

Daugherty, A. 2020. Maria Elvira Salazar defeats Donna Shalala in Florida's 27th Congressional District. *Miami Herald* (Nov. 4). https://www.miamiherald.com/news/politics-government/election/article246867257.html

Diario de Cuba. 2020. Así ven algunos opositores cubanos los resultados de las elecciones en EEUU. (Nov. 11). https://diariodecuba.com/internacional/1604926351_26327.html

diariolasamericas.com. 2020. *de Trump impulsa tema de Los Tres de La Habana.* https://www.diariolasamericas.com/eeuu/campana-trump-impulsa-tema-los-tres-la-habana-n4209065

Entrevista. 2020. Entrevista de Alex Otaola a Trump : "Vas a tener libertad donde viniste, vas a poder ir y regresar". *Cubanos por el Mundo.* (Oct. 17). https://www.youtube.com/watch?v=7rdRyLCxMxM

Fernández, A. 2020. José Daniel Ferrer: Medidas de Biden solo beneficiarían al régimen de Raúl Castro y Díaz-Canel. *CiberCuba* (Nov. 2). https://www.cibercuba.com/noticias/2020-11-02-u207959-e207959-s27061-jose-daniel-ferrer-medidas-biden-solo-beneficiarian-al

Ferrer. 2021. Hija de Ferrer pide libertad para detenidos el 11J: No sabemos qué torturas les están haciendo. *CiberCuba.* (Dec. 1). https://www.cibercuba.com/noticias/2021-08-07-u1-e196568-s27061-hija-ferrer-pide-libertad-detenidos-11j-sabemos-torturas-les

Fowler, III, G., L. Llamas, and L. Cuervo. 2019. Trump ends suspension of the Helms-Burton Act. *National Law Review* (May 3). https://www.natlawreview.com/article/trump-ends-suspension-helms-burton-act

Gaffney, R. 2020. A look into why Florida is a swing-state in presidential races. *WFSU News* (Oct. 3). https://news.wfsu.org/state-news/2020-10-30/a-looking-into-why-florida-is-a-swing-state-in-presidential-races

Geboers, M. and Van De Wiele, C. 2020. Regimes of visibility and the affective affordances of Twitter. *International Journal of Cultural Studies 23*(5):745–765.

Grant, J. 2019. ¿ Qué bola? What's new (and what isn't) in Cuba. *International Journal of Legal Information 47*(3):141–148.

Grenier, G. and Lai, Q. 2020. 2020 FIU Cuba Poll: How Cuban-Americans in Miami View U.S. Policies Toward Cuba. Florida International University. https://cri.fiu.edu/research/cuba-poll/2020-fiu-cuba-poll.pdf

Hansing, K. and Hoffmann, B. 2020. When racial inequalities return: assessing the restratification of Cuban society 60 years after revolution. *Latin American Politics and Society 62*(2):29–52.

Hasell, A. and B. Weeks. 2016. Partisan provocation: The role of partisan news use and emotional responses in political information sharing in social media. *Human Communication Research 42*(4):641–661.

Henry, J.P. 2021. Complicating the role of sight: Photographic methods and visibility in slow violence research. In S. O'Lear (ed.) *A Research Agenda for Geographies of Slow Violence*. London: Edward Elgar.

Herrera, A. and Cañiva, E. 2021. Ghost in the machine: The incompatibility of Cuba's state media monopoly with the existence of independent digital media and the democratization of communication. *Cuba's Digital Revolution: Citizen Innovation and State Policy*. Gainesville, FL: University of Florida Press.

Hui, M. 2020. US Election 2020 Tweets. https://www.kaggle.com/manchunhui/us-election-2020-tweets

Internet. 2021. María Elvira Salazar pide a Biden una estrategia para restaurar internet en Cuba si el régimen corta el servicio el 15N. *CiberCuba*. (Nov. 11). https://www.cibercuba.com/noticias/2021-11-11-u1-e129488-s27061-maria-elvira-salazar-pide-carta-joe-biden-este-listo-apoyar-al

Isla, W. 2021. Primera victoria de la Ley Helms-Burton: Familia expropiada logra acuerdo en demanda contra multinacional. *CiberCuba*. (May 5). https://www.cibercuba.com/noticias/2021-05-27-u199572-e199572-s27061-primera-victoria-ley-helms-burton-familia-estadounidense

Mazzei, P. 2020. In Miami-Dade County, younger Cuban voters offer opening for Trump. *The New York Times*. (Oct. 25). https://www.nytimes.com/2020/10/25/us/miami-cuban-trump-biden.html

Mills, R. 2020. Meet the Cuban YouTuber hoping to turn south Florida into MAGA country. *National Review* (Oct. 5). https://www.nationalreview.com/news/meet-the-cuban-youtuber-hoping-to-turn-south-florida-into-maga-country/

Mueller, M. 2015. A half-hearted romance? A diagnosis and agenda for the relationship between economic geography and actor-network theory (ANT). *Progress in Human Geography 39*(1):65–86.

Multa. 2021. Multa de 3.000 pesos a joven cubano detenido con violencia por grabar cola en Las Tunas (March 4). https://www.cibercuba.com/noticias/2021-03-04-u1-e129488-s27061-multa-3000-pesos-joven-cubano-detenido-violencia-grabar-cola

Papacharissi, Z. 2015. *Affective Publics: Sentiment, Technology, and Politics*. Oxford: Oxford University Press.

Patriotic Union of Cuba. 2018. *Sobre nosotros*. https://www.unpacu.org/en/acerca-de/sobre-unpacu/

Pentón, M. 2020. Caravan participants support Trump, demand liberation of Cuba, Venezuela and Nicaragua. *Miami Herald* (Oct. 11). https://www.miamiherald.com/news/local/community/miami-dade/article246382830.html

Pérez, L. 2003. *Cuba and the United States: Ties of Singular Intimacy*. Athens, GA: University of Georgia Press.

Pink, S. 2013. *Doing Visual Ethnography*. London: Sage.

Rose, G. 2016. *Visual Methodologies: An Introduction to Researching with Visual Materials*. London: Sage.

Ruthhart, B. and Berlin, J. 2020. Campaign trail tracker: Where Trump, Biden and their running mates have traveled in presidential race's final weeks. *Chicago Tribune*. (Nov. 5). https://www.chicagotribune.com/politics/ct-viz-presidential-campaign-trail-tracker-20200917-edspdit2incbfnopchjaelp3uu-htmlstory.html

Sharp, J. 2009. Geography and gender: What belongs to feminist geography? Emotion, power and change. *Progress in Human Geography 33*(1):74–80.

Shaw, I. and B. Warf. 2009. Worlds of affect: Virtual geographies of video games. *Environment and Planning A 41*(6):1332–1343.

Taladrid, S. 2020. How pro-Trump disinformation is swaying a new generation of Cuban-American voters. *The New Yorker* (Oct. 26). https://www.newyorker.com/news/us-journal/how-pro-trump-disinformation-is-swaying-a-new-generation-of-cuban-american-voters

The Borgen Project. 2021. US Support for Las Damas de Blanco in Cuba. https://borgenproject.org/las-damas-de-blanco/

Thrift, N. 2008. *Non-representational Theory: Space, Politics, Affect*. London: Routledge.

U.S. Census Bureau. 2021. *Race and Ethnicity in the United States: 2010 Census and 2020 Census*. Washington, DC: U.S. Census Bureau. https://www.census.gov/library/visualizations/interactive/race-and-ethnicity-in-the-united-state-2010-and-2020-census.html

Vicari, S. 2014. Blogging politics in Cuba: The framing of political discourse in the Cuban blogosphere. *Media, Culture & Society 36*(7):998–1015.

11 "I was Robbed"

Election as Title Match

David Beard and John Heppen

American Political Discourse and Professional Wrestling

Significant work has been done on the interconnectedness of professional wrestling and American political life. The "squared circle" of the wrestling ring has been discussed as a theater where social and political tensions are worked through in a form that can be sold to mass audiences. But more than that, wrestling, as an industry, has produced a significant number of political leaders – perhaps because wrestlers are celebrities and celebrities sometimes enter politics, but perhaps also because wrestlers understand theater in ways that other celebrities do not.

The number of wrestlers to enter politics is huge, although some (like Jerry Lawlor in his run for mayor of Memphis, Tennessee) may have run more as a gimmick than a genuine entry into civic life. Kane (Glenn Jacobs) ran for and won the seat of mayor of Knox County, Tennessee. Jesse Ventura ran for and won a seat as a third-party candidate for governor of Minnesota in 1998. While not a wrestler herself, Linda McMahon ran for the US Senate in Connecticut in 2010 and 2012, losing both times, and later served in Trump's cabinet as Administrator of the Small Business Administration. In the same year that Trump won the presidency, Rhyno (Terrance Guido Gerin) ran as a Republican for the Michigan House of Representatives, in 2016 in his hometown of Dearborn. Rhyno lost, though the point remains – professional wrestlers sometimes become professional politicians.

In this context, especially considering Trump's business relationship and apparent friendship with the McMahon family (who operate the World Wrestling Entertainment business), it is no surprise that Trump entered politics and that Trump campaigned like a wrestler.

The connections between Trump's political rhetoric and the discourses of professional wrestling have been articulated in the popular press and by scholars in Communication, Geography, and Political Science. Works like "How Professional Wrestling Explains American Politics (Especially Donald Trump)" by Oliver Willis (2016) make broad attempts to describe American politics in terms of "faces" and "heels," using the overarching narrative, soap operatic structure of the wrestling match as an analogy for elements of American political life. Heather Bandenburg (2016) tells us that

DOI: 10.4324/9781003260837-11

wrestling and politics both rely on over-the-top characters who fight not just to win but for popularity, using outrageous stunts. Trump's rhetoric is derived in part at least from his background in wrestling. Trump was no stranger to the public before his run for presidency. He hosted the highly rated television reality show (*The Apprentice*) and wrote a bestselling book (*The Art of the Deal*). His life as a multi-millionaire real estate developer in New York put him on the covers of magazines, on television shows, and allowed him to develop real estate properties and hotels in New York and Las Vegas among other places globally. He was a celebrity millionaire before his entry into politics and professional wrestling. He took the name "Trump" and made it into a marketing gimmick for steaks and clothing among other consumer goods.

Trump's background with the professional wrestling industry is expansive and long-lasting. For example, Trump paid a fee to host *Wrestlemania IV* in an attempt to boost attendance at his properties in Atlantic City. Similarly, Trump hosted *Wrestlemania V* in Atlantic City and was featured in an interview during the telecast on Pay Per View (PPV) and closed-circuit television (Margolin 2017). Trump attended *Wrestlemania VII* in Los Angeles, where his future second wife Marla Maples was featured as a guest celebrity. Trump's Trump Taj Mahal in Atlantic City was host to the 1991 World Bodybuilding Federation (WBF) Championship (the WBF was owned by the McMahon family, as well). The success of Wrestlemania, in the first decade, was part of the success of Trump properties.

Trump made several appearances in WWE promotions as a character named "Donald Trump," who becomes part of the fictional world of the WWE. For example, Trump was the main attraction in *Wrestlemania XXIII*, where he engaged in a showdown with Vince McMahon before more than 80,000 spectators. Trump would face McMahon using surrogate wrestlers. Bobby Lashley, a professional wrestler with a background in amateur wrestling, was the stand-in for Trump; Umaga, a Samoan-American wrestler, was the stand-in for WWE owner Vince McMahon. Lashley defeated Umaga and Trump and Lashley proceeded to shave Vince McMahon bald. The event had over 1.2 million PPV buys (a record at the time) and took in a total of over $32 million in revenue (Margolin 2017).

Trump's connection to the WWE is significant enough for him to have been inducted into the WWE Hall of Fame on the night before *Wrestlemania XXIX* at Madison Square Garden. As Bandenburg (2016) noted, "Trump has always been essentially a wrestling gimmick embodied in a real life person."

These connections, traced in the mass media, have been the subject of scholarly reflection, notably in "Smarks, Marks, and the Electorate" by Beard et al. (2020) and in *Donald Trump and the Kayfabe Presidency: Professional Wrestling Rhetoric in the White House* by O'Brien (2020).

Beard et al. (2020) blend a rhetorical analysis of Trump's discourse with the insights of electoral demography to assert a correlation between Trump's rhetorical moves and the rhetorical moves of professional wrestlers, between Trump's popularity (measured by his electoral success) and the popularity of

WWE professional wrestling (measured by attendance at "house shows," traveling wrestling performances in local arenas). Where wrestling is popular, Trump appeared successful. Perhaps, then, "talking like a wrestler" had some electoral advantage in the 2016 election.

Bow O'Brien's monograph (O'Brien 2020) reaches wider, across a broader spectrum, drawing works from political science and political communication to argue both that Trump has learned political theater from the world of professional wrestling and that the administration's decision to exclude his rally speeches from the official record of his activities as president has erased some of the most radical rhetoric from the archives. Our attempt to pull in both tweets and rally speeches is in part inspired by her call to make that presidential discourse part of the record.

Whether we look to scholarly or popular accounts, it's clear that Trump campaigned like a wrestler, used a "wrestling rhetoric," in 2016 and into his presidency. In this chapter, we update these accounts, deepening our understanding of "wrestling rhetoric" through analysis of campaign and post-campaign discourses on Twitter and at rallies.

We'll do this work in two phases. In phase one, we will map the larger patterns of discourse through a quantitative analysis of some 50,000 words of Trumpian tweets and speeches, comparing that map to a sample of pro wrestling "promos" (talks given before a match to build interest or generate heat for the match). Television promos are standard tools in pro wrestling to attract viewers and increase ticket sales. The greater the heat, the tension and anticipation of something big happening is used to generate revenue and viewers. Then, we will engage in some close-reading of examples of post-election Trumpian discourse that exhibits the qualities of the wrestler – specifically, the wrestler who is "building heat" in the audience for a "rematch."

Using Computerized Content Analysis to Explore a Wrestling Rhetoric

For purposes of preliminary statistical analysis, we constructed five corpora. We culled and cleaned two transcripts from Trump speeches after his election loss in 2020: One in Dalton, Georgia on January 4, primarily building enthusiasm for the runoff election), and one rally speech in Washington, D.C. on the January 6 insurrection. We cleaned up the base transcripts found on the website for commercial transcription service, Rev.com. We culled Trump's post-election Tweets from an independent website archiving those Tweets at https://www.thetrumparchive.com/

For comparison, we constructed a small sample corpus of wrestling promos (talks given "to the camera" by professional wrestlers in anticipation of a forthcoming match) by aggregating fan-made transcriptions of promos by Hulk Hogan and by Randy "Macho Man" Savage during their years with the WWF (World Wrestling Federation, now World Wrestling Entertainment).

These corpora were then subject to multidimensional, computational rhetorical analysis using the Software package DICTION 7.0. According to the software's creator, DICTION is

> A multi-platform program written in C++ and Visual C++ that deploys some 10,000 search words in 33 dictionaries. By design, none of the search terms is duplicated from dictionary to dictionary. A "dictionary" is little more than a (carefully constructed) word list. Temporal References (n= 357), for example, consist of such words as adjourns, afternoon, aging, antiquity, autumn, brief, calendar, daily, dawn, etc., while Spatial References (n= 325) include Afghanistan, Argentina, backwoods, borders, bucolic, coast, continent, directions, dislocated, disoriented, distance, etc.
>
> Because DICTION looks at a text from so many different vantage points simultaneously, it provides an unusually comprehensive examination of a given passage. Previous studies with over 40,000 texts show that – generally speaking – these five master variables are completely uncorrelated, ensuring that each provides a completely "fresh" look at any passage being examined.
>
> (Hart and Lind 2011, p. 114)

Using DICTION, then, we submitted roughly 20,000 of Trump's spoken words to this analysis and 30,000 of his words in tweets.

It may be helpful to visualize the ways that DICTION moves from the dictionaries to the five master variables used to describe a text. With a very few exceptions (e.g., word length), DICTION tallies "counts" (appearances of a word listed in a dictionary in a 500-word passage). Then, it adds scores that positively contribute to a rhetorical effect and it subtracts scores that undermine or undercut that rhetorical effect. In the graph below, the dictionaries indicated in square boxes are positively accounted for within the master variable, while the dictionaries indicated in ovals are accounted for negatively, as lowering the overall master variable score.

There are several benefits to using computerized content analysis. As Noor Ghazal Aswad (2019, p. 62) noted, in his analysis of speeches by Clinton and Trump, DICTION

> allows for a mixture of quantitative and qualitative methods (Insch, Moore, and Murphy 1997). Second, because of the standardization of the computer software, it is considered highly reliable and efficient. Third, the software is sensitive to subtleties in a text that even an unbiased and talented coder might not notice
>
> (Bligh, Kohles, and Meindl 2004)

Typically, such software-based analyses would emphasize the distinctions between corpora. For example, Aswad notes that "Donald Trump was significantly more likely to use language emphasizing a shared social identity with

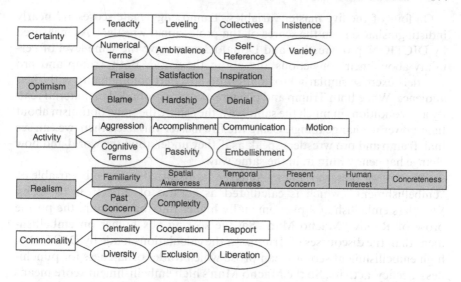

Figure 11.1 Dictionaries and constructed variables for DICTION software.

Full definitions of the master variables are in Appendix 11.1.

his followers and the need to pursue a common goal than Hillary Clinton. Alternately, Hillary Clinton was more likely to use self-referential terms ('I,' 'myself,' 'me') over collective terms ('we,' 'us') when compared to her opponent" (2019, p. 59). While we will note (and where we can, attempt to explain) differences between the language of Trump and the language of professional wrestlers, the commonalities are more significant in supporting our claim that Trump "talks like a wrestler" as a political and public figure.

In Table 11.1, we see aggregate numbers for the five master dimensions of text. These scores represent aggregates achieved through a weighted process (described above) and so don't reflect a single empirical number (the way some of the other categories, like "average word length," do). They are metrics best used to compare against each other and against other texts measured in DICTION.

Table 11.1 Five Master Variables for the Corpora

	Activity	*Optimism*	*Certainty*	*Realism*	*Commonality*
Donald Trump January 6	52.83	46.87	55.81	51.63	50.48
Donald Trump Rally January 4	53.49	48.74	52.95	49.98	49.38
Average of Trump Speeches	53.16	47.81	54.38	50.81	49.93
Trump Post- Election Tweets	51.50	46.44	52.23	45.74	47.88
Hulk Hogan Promos	26.21	50.73	53.53	48.78	49.33
Macho Man Promos	−49.22	33.59	50.00	47.00	50.06
Average of Wrestler Promos	−11.51	42.16	51.77	47.89	49.70

On four of the five master variables, Trump's aggregate scores are nearly indistinguishable from the scores of the professional wrestlers. As measured by DICTION, pro wrestlers and Donald Trump have the same level of certainty about their convictions and about the world. Both Trump and pro wrestlers exercise similarly broad attempts to assert commonality with their audience. While both Trump and pro wrestlers are grounded in material reality and conditions in much the same way, they share the same optimism about their power to change things. Using the five master variables, then, we can see that Trump and pro wrestlers speak as if they understood the world and how change happens within it, in the same way.

(The fifth of these variables, "Activity," is distorted by the sub-variable of "embellishment," which is calculated as a ratio of adjectives to verbs. Wrestlers embellish at a phenomenally higher rate than Trump, the purple prose of Randy "Macho Man" Savage scoring 60x higher on embellishment than the discourses of Trump. In the calculations made by DICTION, high embellishment scores mean prose which trades ornateness for punchiness, agency, activity. So the Macho Man's high embellishment score means that he has a lower overall activity score. Trump appears, in his speeches, to be even more powerful a change agent than pro wrestlers appear to be in theirs.)

At several of the sub-variables, too, the lexical choices of Trump are also indistinguishable from our professional wrestling sample. For example, on the dimension of "complexity," or average number of characters-per-word in the text, Trump and the professional wrestlers hover at about 4.3 characters per word in any 500-word sample. (Trump's tweets are slightly longer, but note that the software calls hashtags which agglutinate words, like "#stopthe-steal," as one word.) The complexity dimension in DICTION asserts, following Rudolph Flesch's work on "reading level," that convoluted phrasings make a text's ideas unclear. Trump and the wrestlers are plenty clear, on this measure (Table 11.2).

In the DICTION dimension of "Leveling Terms," or "words used to ignore individual differences and to build a sense of completeness and assurance," Trump and the professional wrestlers are nearly indistinguishable. To ascertain the score in "Leveling Terms," DICTION counts totalizing terms

Table 11.2 Complexity Subscore is Indistinguishable across Trump and Professional Wrestlers

	Complexity
Donald Trump January 6	4.33
Donald Trump Rally January 4	4.39
Average of Trump Speeches	4.36
Trump Post-Election Tweets	4.66
Hulk Hogan Promos	4.32
Macho Man Promos	4.13
Average of Wrestler Promos	4.23

Table 11.3 Levelling Terms Subscore is Indistinguishable
across Trump and Professional Wrestlers

	Leveling Terms
Donald Trump January 6	109.42
Donald Trump Rally January 4	98.88
Average of Trump Speeches	104.15
Trump Post-Election Tweets.docx.dfxml	102.08
Hulk Hogan Promos	105.97
Macho Man Promos	110.07
Average of Wrestler Promos	108.02

(everybody, anyone, each, fully), adverbs of permanence (always, completely, inevitably, consistently), and resolute adjectives (unconditional, consummate, absolute, open-and-shut). Trump builds a sense of security in his audience the same way that wrestlers build the energy, the confidence, of their fans – by placing everything in their hands, under their control, the way it has always been (Table 11.3).

In the DICTION dimension of "Centrality" (which tracks "substantive agreement on core values," Trump builds energy with his audience by embodying the values they already share, in the ways that professional wrestlers embody the values of the fans. To calculate Centrality, DICTION counts "indigenous terms (native, basic, innate) and designations of legitimacy (orthodox, decorum, constitutional, ratified), systematicity (paradigm, bureaucratic, ritualistic), and typicality (standardized, matter-of-fact, regularity). Also included are terms of congruence (conformity, mandate, unanimous), predictability (expected, continuity, reliable), and universality (womankind, perennial, landmarks)." Trump appeals to the values of his voters, waving their signs at a rally, the way the Hulk Hogan or the Hulkster appeals to the values of the Hulkamaniacs (Table 11.4).

In terms of DICTION's dimension of "Exclusion," or a measure of social isolation – independence (whether achieved voluntarily, in the spirit of rugged individualism, or involuntary, in a kind of ostracism from mainstream

Table 11.4 Centrality Subscore is Indistinguishable across
Trump and Professional Wrestlers

	Centrality
Donald Trump January 6	14.85
Donald Trump Rally January 4	8.61
Average of Trump Speeches	11.73
Trump Post-Election Tweets.docx.dfxml	10.60
Hulk Hogan Promos	9.26
Macho Man Promos	11.45
Average of Wrestler Promos	10.36

Table 11.5 Exclusion Subscore is Indistinguishable across Trump and Professional Wrestlers

	Exclusion
Donald Trump January 6	42.47
Donald Trump Rally January 4	35.29
Average of Trump Speeches	38.88
Trump Post-Election Tweets.docx.dfxml	36.73
Hulk Hogan Promos	38.82
Macho Man Promos	39.56
Average of Wrestler Promos	39.19

life) – Trump and the sample of professional wrestlers invoke the same kind of rugged individualism with nearly the same frequency – roughly 38–39 words setting this tone in every 500-word sample of text (Table 11.5).

In the aggregate, then, the overall rhetorical dimensions (revealed through lexical analysis) of Trump and of the professional wrestlers substantially overlap: sharing a baseline worldview, expressed clearly in values shared between speaker and audience as rugged individuals, certain of the inevitability of their triumph.

With that said, it would be useful to explore just a few of the differences between Trump's discourse and that of the professional wrestlers. Some may be statistical errors generated by the limitations of the software. For example, DICTION attempts to trace self-reference in a text. DICTION traces "Self-Reference" by counting all first-person references in a typical 500-word passage of text, including *I, I'd, I'll, I'm, I've, me, mine, my, m*yself. According to the DICTION manual, "Self- references are treated as acts of indexing whereby the locus of action appears to reside in the speaker and not in the world at large" – and Trump appears to score more strongly in this area, in his speeches sampled, the average score was 5.19. Across the wrestling promos analyzed, the average score was 2.66. This analysis may be limited by the software, however, as it fails to account for references to the self in the third person – a rhetorical strategy deployed by "The Hulk" and "The Macho Man" on a regular basis. Wrestlers' scores in the self-reference category, then, may be deflated because they do not count references to the self in the third person alongside the first-person references. Further analysis is called for.

Others reflect genuine differences in the rhetorical style of the wrestlers. Trump's embellishment score, noted above, in spoken discourse is less than .5 (as determined by a ratio of adjectives to verbs). Randy Macho Man Savage scores a 30 – more than 60 times higher a score than Trump. Oddly, perhaps, the professional wrestler uses more adjectives, more flowery language, than the former president of the United States.

There is one data point which is radically erratic, an outlier, even within the context of Trump's language use. Between January 4 and January 6, 2021, Trump's "insistence" score leaps from 30 to 70. "Insistence," within DICTION,

Table 11.6 Insistence Subscore Ratchets Up, for Trump, on January 6

	Insistence
Donald Trump January 6	70.41
Donald Trump Rally January 4	30.64
Average of Trump Speeches	50.53
Trump Post-Election Tweets	34.30
Hulk Hogan Promos	35.97
Macho Man Promos	26.75
Average of Wrestler Promos	31.36

measures repetition of key terms. To calculate an insistence score, DICTION singles out all nouns or noun-related adjectives used three or more times (in a 500-word segment of text). When the score is high, indicating a large dependence on the same, repeated key terms, the score is taken to indicate a preference for a limited, ordered world (Table 11.6).

Over the 30,000 words found in the Tweets in the 60-day period between losing the election and the insurrection, Trump's insistence score is relatively low – only marginally higher than the professional wrestlers' scores (34.30 to 31.36). On January 4, his insistence score for the rally speech is in the same ballpark: 30.64. On January 6, his insistence score more than doubled, to 70.41, for the speech at the start of the insurrection.

At this point, conjecture on this score difference would be only that, conjecture. Working on the metaphor we are establishing here, in which Trumpian discourse is similar to wrestling discourse, and that Trump uses language to "build heat" the way wrestlers do, Trump is no longer building heat on January 6 – he is starting what he hopes will be the final match. Perhaps that turn increases the insistence score for the text.

Let's take a closer look at how Trump "builds heat" in the debates and in his speeches between the loss of the 2020 election, the Georgia runoff, and the January 6 insurrection.

Using Close Reading to Explore a Wrestling Rhetoric

Wrestling is built upon narratives that rarely reach closure: in the lead-up to a title match, wrestlers engage in promo segments that "build heat." Heat leads an audience to see the match, but the title match resolution is often not definitive. One wrestler may insist that they were "robbed" of their title, claiming that the other wrestler "cheated" (regardless of whether they did), "building heat," again, for a rematch, with hopefully even bigger ratings than the first match.

We believe that through the 2020 election and after, Trump spoke like a wrestler "building heat" prior to a match and, after losing the title bout, building heat for a rematch or an event like January 6, 2021.

Table 11.7 Locations of the 2020 Election Debates

Tuesday, September 29, 2020	Case Western Reserve University	Cleveland, Ohio	Chris Wallace
Thursday, October 15, 2020	Arsht Center	Miami, Florida	Steve Scully
Thursday, October 22, 2020	Belmont University	Nashville, Tennessee	Kristen Welker

Phase One: "Building Heat" and the 2020 Debates

Trump's Summer and Fall 2020 rallies, and the debates with Joe Biden, were structured to "build heat" the way that house shows "build heat" for a forthcoming title match. Trump approached the debates with Joe Biden as "promo" segments, opportunities to speak to the audience, not to Biden, to build their excitement for the match (the election) (Table 11.7).

Just before the first debate, while at a rally in Middletown, PA on September 26, 2020, Trump paints himself as the underdog: "He's got 47 years. I've got three and a half years, so we'll see. He's got 47 years of experience" ("Middletown, PA Rally Speech Transcript Sept. 26"). As the underdog, he's taking on the big man: "We've spent the last four years reversing the damage Joe Biden inflicted over his 47 years in politics" ("Middletown, PA Rally Speech"). (Of course, the incumbent is rarely the underdog, any more than the champion is the underdog to the challenger in the ring. But Trump activates the underdog rhetoric to galvanize the crowd, as a wrestler might.)

Building heat, Trump pokes at Biden relentlessly, calling him "a low-energy individual" and, more baldly insulting, "sleepy Joe" ("Middletown, PA Rally Speech"). Via Twitter, on the 27th, Trump keeps poking:

> Sep 27 2020 – 9:34:08 AM EST
> I will be strongly demanding a Drug Test of Sleepy Joe Biden prior to, or after, the Debate on Tuesday night. Naturally, I will agree to take one also. His Debate performances have been record-setting UNEVEN, to put it mildly. Only drugs could have caused this discrepancy???

Combative, insulting – this is the talk of the wrestler.

The night after the first debate, on Tuesday, September 29, 2020, at Case Western Reserve University in Cleveland, Ohio (with moderator Chris Wallace), Trump flies to Duluth, MN for a rally. There, Trump makes sure that the audience knows that he's a ratings winner, that his contest, the night before, was both a victory as a debate and a victory in the ratings. The first victory is so very important to Trump, who measures his success by his popularity and media draw:

> The verdict is in and they say that we, we, all of us, won big last night… In the history of cable television, it had the highest ratings of any show in the history of cable television. It had the second highest ratings of

overall television in the history of television. Does anybody know what was first? Like M*A*S*H or something? I guess M*A*S*H, they had the final episode of M*A*S*H, and I don't know what was first. Does anybody know? But we were second in the history of all of television, but the biggest ratings in the history of cable television. It's an honor.

(Duluth, Minnesota Campaign Rally Transcript September 30)

While these statements are inaccurate (Trump also scored lower ratings than several Superbowl broadcasts, the M*A*S*H* finale is only #1 in the scripted drama category).

Similarly, Trump frames the size of the crowd at the Duluth event as another victory over Biden:

Look at this crowd, this was supposed to be a few people. This was supposed to be just a little celebration. I said, "Oh, good. A little celebration." I figured a couple of hundred people like Sleepy Joe gets it his best day. I said, "How many people?" "About 10,000, Sir." I said, "Oh, that's nice"

(Duluth, Minnesota Campaign Rally Transcript).

Trump is building heat, competing with Biden in every way he can. After all, he claims, "I did more in 47 months than sleepy Joe Biden did in 47 years" (Duluth, Minnesota Campaign Rally Transcript). The message was reiterated on Twitter, on October 1, 2020 at 11:14:28 AM EST: "I won the debate big, based on compilation of polls etc. Thank you!"

Setting up the underdog narrative, Trump emphasizes that he overcomes a rigged system to achieve this victory. In his own words, describing the behavior of Fox News anchor Chris Wallace, Trump claims that "I was debating two people last night, I was debating two people last night… Well, well, well, thank Chris Wallace, 'Okay. let's go to the next question. He's in trouble. He's in trouble. Let me protect him for Fox'" (Duluth, Minnesota Campaign Rally Transcript). He's also setting up the post-election loss narrative, which is a classic wrestling narrative about being "robbed."

Phase Two: "I was Robbed" after the 2020 Election Loss

After his eventual electoral defeat, as states are called for Biden by major media outlets, Trump enters the phase of wrestling discourse where he claims to have been robbed, building even more heat for the "rematch" – in rallies where he tried to bring voters out for the special election in Georgia or in the rally on January 6. The post-election rallies are noted in Table 11.8.

For our purposes, we will look at the speech on January 6. This rally speech crystallized the rhetoric of being robbed. On the Ellipse, Trump claimed that the election was rigged: "they rigged an election. They rigged it like they've never rigged an election before." The agents, the nefarious ones, are the Democrats and the media. Trump claims that "All of us here today do not

Table 11.8 Trump's Post-Election Rally Locations

December 5, 2020	Valdosta Regional Airport, Valdosta, GA	David Perdue, Kelly Loeffler, Vernon Jones, Burt Jones (pledge reciter), Gary Black (singer), Bubba McDonald
January 4, 2021	Dalton Municipal Airport, Dalton, GA	Kelly Loeffler, Marjorie Taylor Greene, Ivanka Trump, Donald Trump Jr., Kimberly Guilfoyle
January 6, 2021	The Ellipse, Washington, DC	Rudy Giuliani, Mo Brooks, Madison Cawthorn, John C. Eastman, Eric Trump, Lara Trump, Kimberly Guilfoyle, Donald Trump Jr.

want to see our election victory stolen by emboldened radical left Democrats, which is what they're doing and stolen by the fake news media."

Continuing the rhetorical assertions he made before the election, he claimed victory: "I've been in two elections. I won them both and the second one, I won much bigger than the first … We didn't lose." The Champ, having lost the title, insists that he didn't lose, or if he did lose, the match was rigged and the win was stolen.

"I was robbed" is a common refrain or outburst in professional wrestling to generate heat or maintain excitement in a storyline or individual wrestler. Typically, the wrestler was robbed due to cheating by an opponent that was missed or overlooked by the referee due to incompetence or outside distraction. (In some cases, the referee, "under bribery" or orders by the promoters, allows the cheating to happen.)

In most cases, the chicanery is scripted, and so both wrestlers know that the "robbery" is part of the theater. Under some circumstances, the chicanery occurs without the knowledge of one of the wrestling parties. The most famous of these occurrences was "The Montreal Screwjob." This took place in an event between Bret "Hitman" Hart and Shawn Michaels on November 9, 1997 in Montreal, QC, Canada as the main event in the "Survivor Series Pay Per View". Hart (a wrestler in the famous Hart family from Calgary Alberta) was one of the most popular stars in professional wrestling. Hart's match with Michaels was to be his last match as he had signed a contract to go to the rival World Championship Wrestling (WCW) after this match. In the predetermined world of professional wrestling, Bret Hart was not supposed to lose against Michaels and give up his title later as Vince McMahon, Hart and Michaels agreed to have the match end in a disqualification and Hart was supposed to lose the title later (Foley 2000). In professional wrestling a disqualification does not result in a title change. But the match ended with a submission hold as referee Earl Hebner and Vince McMahon ordered the timekeeper to ring the bell while Michaels held Hart in a submission hold. Hart had not submitted, in clear view of everyone. Michaels was declared the victor and an irate crowd roared in disapproval and Hart appeared to be furious.

A screwjob, like the Montreal Screwjob, is an element of professional wrestling storytelling used to build interest and either sympathy for the victim or hatred for the perpetrator among the fans. These kinds of screw-jobs, whether fully scripted or not, become part of a storyline of revenge and redemption. When Trump invokes, with the wrestler's language and rhetoric, claims that he has been robbed, that the ref (the county canvassers, the state certification boards, perhaps, or the Senate) blew the call, he is perhaps invoking the revenge and redemption narrative for his campaign, too.

Significance: Understanding Wresting Rhetorics in the Past and in the Future

In this chapter, we have attempted, through both qualitative and quantitative methods, to assert similarities between the rhetoric of professional wrestling and the rhetoric of Donald Trump. Such analogies, to be useful, do not need to function in full recognition by the audience. You don't need to be a wrestling fan to be persuaded by wrestling rhetorics. As Hart and Lind (2013) note, in his explanation of the value of quantitative analysis: "word choices often aggregate into patterns of meaning for audiences who only half-notice the messages they consume each day; and patterns of dissociated words accumulate in audiences' minds to form overall impressions that can be highly consequential" (p. 115). An audience does not need to recognize that Trump's communication style mirrors that of a professional wrestler for that communication style to nonetheless work upon them, in the way that wrestlers' promos work the emotions of their audiences.

But – aware of his rhetorical choices or not, we were all "marks" for Trump. He won an election and won more votes than any incumbent has won before. His rhetorical flourishes worked on a lot of people, including former Obama voters who were drawn to his talk of shared values and use of leveling terms to bind his audience into a common project. His rhetorical flourishes appealed to the rugged individualist who values liberty and individual success and can see themselves in the champion who stirs the Hulkamaniacs to unleash their own power.

The story he told, after the election, was a wrestling story. Wrestling is built upon narratives that rarely reach closure. The title match may come, but, often, its resolution is not definitive. One wrestler may insist that they were "robbed" of their title, claiming that the other wrestler "cheated" (regardless of whether they did), "building heat," again, for a rematch, with hopefully even bigger ratings than the first match.

Perhaps Trump believed that January 6 was to be the rematch, and the belt would fall into his hands again. (The increase in his "insistence" score tells us something about a difference in the rhetorical choices he made in address, at least.) More likely, the rematch will be in 2024, and our analyses may both explain the past rhetoric of the 45th president as well as prepare us to understand his strategies in the future.

Appendix

Appendix 11.1 Five Master Variables in DICTION

Master Variable	Operationalization within the DICTION software package.
Certainty	Certainty includes words indicating resoluteness, inflexibility, and completeness, and its raw scores are calculated by summing word counts in the tenacity, leveling, collectives, and insistence dictionaries and then subtracting the numerical terms, ambivalence, self-reference, and variety dictionaries' word counts.
Optimism	Optimism refers to language endorsing some person, group, concept, or event, or highlighting positive entailments. Optimism's raw scores are calculated by summing word counts in the praise, satisfaction, and inspiration dictionaries and then subtracting blame, hardship, and denial dictionaries' word counts.
Activity	Activity refers to movement, change, the implementation of ideas, and the avoidance of inertia. Activity's raw scores are calculated by summing word counts in the aggression, accomplishment, communication, and motion dictionaries and then subtracting word counts in the cognitive terms, embellishment, and passivity dictionaries.
Realism	Realism refers to words associated with tangible, immediate, recognizable matters that affect people's everyday lives. Realism's raw scores are calculated by summing word counts in the familiarity, spatial awareness, temporal awareness, present concern, human interest, and concreteness dictionaries and then subtracting word counts in the past concern and complexity dictionaries
Commonality	Commonality measures language associated with human communitarian instincts that emphasizes the agreed-upon values of a group and rejecting idiosyncratic modes of engagement. Commonality's raw scores are calculated by summing word counts in the centrality, cooperation, and rapport dictionaries and subtracting word counts in the diversity, exclusion, and liberation dictionaries.

From Hart and Carroll (2013)

References

Bandenburg, H. 2016. Did the WWE and wrestling help get Donald Trump elected? *The Independent* (Nov. 14). https://www.independent.co.uk/sport/general/wwe-mma-wrestling/donald-trump-wwe-wrestling-us-election-latest-policies-president-elect-a7416901.html

Beard, D., Heppen, J., and Millett, M. 2020. Smarks, marks, and the electorate: Trump, wrestling rhetorics, and electoral politics. In B. Warf (ed.) *Political Landscapes of Donald Trump*. pp. 193–213. London: Routledge.

Foley, M. 2000. *Have a Nice Day. A Tale of Blood and Sweatsocks.* New York: Regan Books.

Ghazal, A. 2019. Exploring charismatic leadership: A comparative analysis of the rhetoric of Hillary Clinton and Donald Trump in the 2016 presidential election. *Presidential Studies Quarterly* 49(1):56–74.

Hart, R. and Carroll, C. 2013. *DICTION 7.0 Help Manual*. Austin, TX: Digitext.

Hart, R. and Lind, C. 2011 The rhetoric of Islamic activism: A DICTION study. *Dynamics of Asymmetric Conflict* 4(2):113–125.

Hart, R., and Lind, C. 2013. "Walking the Partisan Line: Mitt Romney in the 2012 Campaign." In APSA 2013 *Annual Meeting Paper, American Political Science Association 2013 Annual Meeting*.

Margolin, L. 2017. *TrumpMania: Vince McMahon, WWE and the Making of America's 45th President*. New York: H. Delilah Business and Career Press

O'Brien, S. 2020. *Donald Trump and the Kayfabe Presidency: Professional Wrestling Rhetoric in the White House*. London: Palgrave Macmillan.

Willis, O. 2016. How professional wrestling explains American politics (especially Donald Trump). *Medium* (March 2). https://medium.com/@owillis/how-professional-wrestling-explains-american-politics-especially-donald-trump-5449df1db9de

Index

Taylor & Francis Group
an **informa** business

Taylor & Francis eBooks

www.taylorfrancis.com

A single destination for eBooks from Taylor & Francis
with increased functionality and an improved user
experience to meet the needs of our customers.

90,000+ eBooks of award-winning academic content in
Humanities, Social Science, Science, Technology, Engineering,
and Medical written by a global network of editors and authors.

TAYLOR & FRANCIS EBOOKS OFFERS:

A streamlined
experience for
our library
customers

A single point
of discovery
for all of our
eBook content

Improved
search and
discovery of
content at both
book and
chapter level

REQUEST A FREE TRIAL
support@taylorfrancis.com

 Routledge
Taylor & Francis Group

 CRC Press
Taylor & Francis Group

Printed in the United States
by Baker & Taylor Publisher Services

Printed in the United States
by Baker & Taylor Publisher Services